Dielectric Relaxation

Dielectric Relaxation

VERA V. DANIEL
Electrical Research Association
Leatherhead, Surrey, England

1967
ACADEMIC PRESS
LONDON AND NEW YORK

ACADEMIC PRESS INC. (LONDON) LTD
BERKELEY SQUARE HOUSE
BERKELEY SQUARE
LONDON, W.1.

U.S. Edition published by
ACADEMIC PRESS INC.
111 FIFTH AVENUE
NEW YORK, NEW YORK 10003

Copyright © 1967 by ACADEMIC PRESS INC. (LONDON) LTD

All Rights Reserved
No part of this book may be reproduced in any form by photostat, microfilm, or any other means, without written permission from the publishers

Library of Congress Catalog Card Number: 67–28004

Printed in Great Britain by
William Clowes and Sons Ltd., London and Beccles

"The race is not to the swift nor the battle to the strong—but time and chance happeneth to them all."
Ecclesiastes, Chapter 9, Verse 11.

Preface

This monograph forms part of the work of the Electrical Research Association. One of the functions of this organization is to advise industry on the use of insulation. The monograph aims to explain theoretical ideas as simply as possible and to relate them to the complicated systems in which dielectric relaxation takes place.

I am indebted to several present and former colleagues for helpful discussions, Mr C. G. Garton in particular. The treatment of electrostatic interaction as positive feedback was suggested by Dr (now Professor) L. E. Cross, who also commented on Chapter 16. Dr H. Pelzer helped with Chapter 5 and supplied Appendix IV. Thanks are also due to Professor H. Fröhlich and Dr B. K. P. Scaife for helpful discussions. Dr Scaife pointed out several qualifications regarding electrostatic interaction. Dr. R. Fürth kindly read Chapter 4 and provided helpful comments on stochastic processes. Correspondence with Professor R. E. Burgess, Dr. A. C. Lynch, Dr R. J. Meakins and Dr B. Szigeti contributed to the clarification of certain points. Last, but not least, I should like to thank Mrs E. Denny for her flawless typing, her skill in decoding and her unfailing patience.

Electrical Research Association
June 1967

VERA DANIEL

Contents

PREFACE vii
INTRODUCTION xiii

1. Relaxation in electrical circuits
   A. Relaxation times defined by differential equations 2
   B. The differential equation for the simplest resonant circuit . . 10
   References 12

2. Relaxation in dielectrics with a single relaxation time
   A. Macroscopic derivation of the Debye equations 13
   B. Molecular models with a single relaxation time 19
   C. An idealized treatment of electrostatic interaction 27
   References 31

3. The thermodynamics of relaxation
   A. Equilibrium considerations 33
   B. Relaxation in terms of irreversible thermodynamics 39
   C. The dissipation of energy 42
   References 45

4. Fluctuations and dielectric relaxation
   A. Equilibrium and fluctuations in a microscopic assembly . . . 47
   B. Fluctuations in the bistable model in equilibrium 49
   C. The rate of fluctuations 53
   D. The statistical significance of the relaxation time for the bistable model 56
   E. The physical significance of the relaxation time and the theory of rate processes 61
   References 64

5. Systems with many relaxation times
   A. The experimental evidence 65
   B. The superposition of delayed responses; decay functions . . . 66
   C. Integral formulations of the principle of superposition, and the distribution function 69
   D. The complex dielectric constant 72
   E. An ambiguity concerning the distribution function: retardation and relaxation 73
   F. The Kramers–Kronig relations 75
   G. An historical note 76
   References 76

6. Dielectric measurements and their interpretation
  A. Step-function measurements 79
  B. Alternating current measurements for frequencies below 10^7 c/s . 87
  C. Alternating current measurements for frequencies above 1 Mc/s . 91
  D. Electrode assemblies 92
  References 93

7. Empirical methods for the evaluation of dielectric measurements
  A. Methods of displaying relaxation peaks 95
  B. Distributions of relaxation times 99
  C. The interpretation of wide loss peaks in terms of activation energies 105
  References 109

8. Electrostatic interaction
  A. The electric field inside cavities in a continuous dielectric . . . 111
  B. The static dielectric constant for a real dielectric 113
  C. Relaxation in the presence of electrostatic interaction 117
  References 121

9. Resonance absorption
  A. Microwave absorption in gases 123
  B. The broadening of spectral lines in gases 124
  C. Resonance absorption in solids 128
  D. The dielectric and optical constants 129
  E. The transition from relaxation to resonance 130
  References 132

10. Dielectric relaxation in gases. 135

11. Relaxation in liquids
  A. A classification of liquids 144
  B. The relaxation of dipole orientation in liquids of Group II . . 145
  C. Associated liquids 154
  D. Liquids of very high viscosity 159
  References 162

12. Relaxation in crystalline solids
  A. Dipolar "rotation" in molecular solids 165
  B. Ionic crystals 176
  References 182

13. Insulating materials with amorphous and glassy constituents
    A. Organic polymers 185
    B. Inorganic glasses 199
    References 200

14. Heterogeneous dielectrics, non-ohmic conduction and electrode effects
    A. Heterogeneous mixtures with linear response: the Maxwell–Wagner effect 203
    B. Field dependent dielectric loss in thin liquid films 208
    C. Electrode effects 211
    References 216

15. Hydrogen bonded solids
    A. Secondary long-chain alcohols 217
    B. Ice 221
    C. The role of hydrogen bonding in practical insulation 228
    References 231

16. Ferroelectrics
    A. Introduction 233
    B. Experimental criteria of ferroelectricity 233
    C. Ferroelectricity and order–disorder transitions 236
    D. The thermodynamic theory of order–disorder transitions . . . 236
    E. Fluctuations near the critical temperature and the role of field theory 242
    F. Relaxation in triglycine sulphate 244
    G. The perovskite group and lattice dynamics 246
    H. Relaxation by movements of domain boundaries 250
    I. Relaxation ferroelectrics 250
    References 253

APPENDICES
  I. Units and constants 255
  II. Formulae for the capacitance and equivalent circuits . . 259
  III. The limiting values of ε_s and tan δ for a given number of dipoles 263
  IV. The four forms of the Duhamel integral 265
  V. Some general expressions relating to distribution functions 267
  VI. Abstracts and collections of data 271

Introduction

Relaxation is a process. The term applies, strictly speaking, to linear systems where a response and a stimulus are proportional to one another in equilibrium. Relaxation is a delayed response to a changing stimulus in such a system. Dielectric relaxation occurs in dielectrics, that is in insulating materials with negligible or small electrical conductivity. The stimulus is almost always an electrical field, the response a polarization. The time lag between field and polarization implies an irreversible degradation of free energy to heat.

The objective of this monograph is to describe and interpret the time dependence of the electrical response of dielectrics. Interpretation is difficult because the observable relationship between polarization and field is simple in the cases relevant for dielectric relaxation and because the measurements have relatively little information content. The response of the dielectric can be described by a set of linear differential equations and many models can be devised which correspond to the same differential equations. When the dielectric relaxation of a given material has been measured the investigator is in the position of a man presented with a black box which has two terminals. He may apply alternating fields of various kinds and he may heat the box but he is not allowed to look inside, and he finds that the box responds as if it contained a combination of capacitors and resistors.

The reaction of our investigator to the puzzle presented by the black box will differ according to whether he is a mathematician, electrical engineer, physicist or chemist. The mathematician will be satisfied by a description in terms of differential equations and the engineer by an equivalent circuit. However, the physicist or chemist will want an interpretation in terms of the structure of the material whose response can be represented by the black box. The materials scientists will often be disappointed.

The relaxation of the polarization in response to a change of the field applied to a material is not due directly to the pull of the field, as suggested by the naive imagination. It is brought about by thermal motion, and fields of the magnitude relevant to dielectric relaxation perturb the thermal motion only slightly.

The structural interpretation of dielectric relaxation is a difficult problem in statistical thermodynamics. It can for many materials be approached by considering dipoles of molecular size whose orientation or magnitude fluctuates spontaneously, in thermal motion. The dielectric constant of the material as a whole is arrived at by way of these fluctuations but the theory is very difficult because of the electrostatic interaction between dipoles. In some ionic crystals the analysis in terms of dipoles is less fruitful than an analysis in terms of thermal vibrations. This also is a theoretically difficult task forming part of lattice dynamics. In still other materials relaxation is due to electrical conduction over paths of limited length. Here dielectric relaxation borders on semiconductor physics.

The theories involved in dielectric relaxation are presented as simply as possible, often in terms of simplified models. In particular, the "bistable" model is treated quantitatively from several viewpoints. By this means it is hoped to lessen the confusion which often arises where a theoretical approach is difficult. Theoreticians necessarily use exact mathematical language to describe models and state the limitations of the validity of their models in this language, which experimenters find difficult. This does no harm where the limits of validity are wide, but in dielectric relaxation the validity of assumptions holds only within narrow limits. The experimenter needs to understand the limits of validity of the model he uses, since he has to keep them constantly in view.

While theory is concentrated mainly in the first half of the monograph the second half aims to give a systematic survey of mechanisms of relaxation in materials of theoretical and practical interest. The monograph is not a complete review of the literature of dielectric relaxation; such a review would have demanded a much longer book. Apologies are due to authors whose relevant publications are not quoted.

Mechanisms of dielectric relaxation are often subtle and complicated, as in ice and the ferroelectrics. Most of the cases where a satisfactory physical interpretation has been achieved have in common that dielectric measurements were not used in isolation but were complemented by other physical or chemical techniques. Only in this way was it possible to decide between the multiplicity of physical models which might satisfy the dielectric data. Dielectric relaxation as a subject is neither self-contained nor self-checking.

CHAPTER 1

Relaxation in Electrical Circuits

When an insulating material is provided with electrodes, and a time-dependent voltage $V(t)$ is applied to these electrodes, the charge Q on them changes with time, and a current I flows. Dielectric relaxation is concerned with the timing of this electrical response, that is with the relationships of $V(t)$, $Q(t)$ and $I(t)$ for different materials. It is a feature of dielectric relaxation that these relationships obey the same mathematical formalism as the relationship of $V(t)$, $Q(t)$ and $I(t)$ for electrical circuits consisting of capacitances, resistances and inductances, that is passive electrical networks. Certain passive networks are mathematical analogues of certain dielectric materials. This is confusing since any measurement of $V(t)$, $Q(t)$ and $I(t)$ needs to be carried out with the aid of some electrical circuit.

This chapter is concerned with circuits, on the basis of the elementary theory of electricity[1,2]. It will be shown that the response of passive networks to changing voltages may be described by linear differential equations and that constant relaxation times τ may be defined which characterize a given process in a given circuit. Only elementary mathematical tools will be used, namely elementary calculus and complex variables.

Analogue circuits for dielectrics mostly contain only capacitances C and resistances R. A capacitor consists of two ideal electrodes separated by vacuum or by some insulating material. If a voltage V is applied to a capacitor it holds a charge Q

$$Q = CV. \qquad (1.1)$$

The capacitance C is defined as

$$C = \varepsilon C_0 \qquad (1.2)$$

where C_0 is a constant determined by the geometry of the capacitor while ε is a material constant, called dielectric constant by physicists and chemists and relative permittivity (denoted by the symbol κ) by electrical engineers. In this monograph the physicists' terminology will mostly be used in conjunction with the c.g.s. system (see Appendices I and II).

Dielectric relaxation is concerned with the fact that for any material with which a capacitor may be filled, ε is variable. In other words, the capacitance of a real capacitor changes when a time-dependent voltage is applied to it. However, in circuit theory one makes the assumption

$$\frac{dC}{dt} \equiv 0 \qquad (1.3)$$

and allows for the resultant inaccuracy by the introduction of fictitious circuit elements. More will be said about this later, while at present we shall treat circuits with pure capacitances characterized by equation (1.3).

A. RELAXATION TIMES DEFINED BY DIFFERENTIAL EQUATIONS

One of the simplest possible circuits consists of a capacitor C_s in series with a resistor; the latter has a constant resistance R_s which is defined by Ohm's law. Figure 1 shows this circuit with terminals to which an external voltage may be applied.

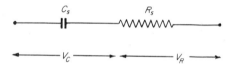

FIG. 1. Series combination of a capacitor and a resistor.

The time-dependent behaviour of the circuit in Fig. 1 may be derived as follows: elementary circuit[1,2] theory demands that the current through elements in series be continuous, and that the voltage across the terminals should be the sum of the voltages across the two circuit elements C_s and R_s.

$$V = V_C + V_R. \qquad (1.4)$$

When V varies with time, V_C and V_R vary also. Since the charge on the capacitor is given by equation (1.1) and condition (1.3) holds, we have for any change of V_C a change of charge

$$\frac{dQ}{dt} = C_s \cdot \frac{dV_C}{dt} \qquad (1.5)$$

while

$$I = \frac{dQ}{dt} \qquad (1.6)$$

is the definition of the current (in this case a pure displacement current) through the capacitor. Since the current is continuous it flows through the resistor and according to Ohm's law the voltage across the resistor is

$$V_R = IR_s \qquad (1.7)$$

1. RELAXATION IN ELECTRICAL CIRCUITS

while according to equation (1.1) the voltage across the capacitor is $V_C = Q/C_s$.

When the voltages are added according to equation (1.4) we have

$$IR_s + \frac{Q}{C_s} = V. \tag{1.8}$$

In view of equation (1.6) this constitutes a differential equation for the charge Q

$$R_s \frac{dQ}{dt} + \frac{Q}{C_s} = V$$

which may be multiplied by C_s and written

$$\tau \frac{dQ}{dt} + Q = C_s V \tag{1.9}$$

where

$$\tau = C_s R_s \tag{1.10}$$

is a constant of the dimension of a time.

Equation (1.9) is a linear differential equation and may be integrated for a given $V(t)$. Such equations are basic to circuit theory and their treatment is discussed in detail by Guillemin[2]. We shall here quote only two solutions for special cases of $V(t)$. For a so-called step-function in voltage, where $V = 0$ for $t \leq 0$ and $V = V_0$ for $t > 0$, the integral of equation (1.9) is

$$Q = C_s V_0 (1 - e^{-\frac{t}{\tau}}) \tag{1.11}$$

which means that Q rises exponentially from $Q = 0$ at $t = 0$ and that it reaches a fraction $1/e$ of its final value after a time τ. This shows the significance of τ as a delay time. In physics and chemistry τ is called a "relaxation time", in electrical engineering a "time constant".

For the discharging of the capacitor, that is for a step-function $V = V_0$ for $t < 0$ and $V = 0$ for $t > 0$, we have

$$Q = C_s V_0 \, e^{-\frac{t}{\tau}}. \tag{1.12}$$

For a sinusoidal applied voltage

$$V(t) = V_0 \cos \omega t \tag{1.13}$$

where $\omega = 2\pi f$ is the circular frequency, the differential equation (1.9) may be solved using complex variables. In doing so, we assume a complex quantity† for the voltage

$$V^*(t) = V_0 \, e^{i\omega t} = V_0 (\cos \omega t + i \sin \omega t) \tag{1.14}$$

† The use of an asterisk to denote a complex quantity is not customary in mathematics but has become so common in the literature of dielectric relaxation that it will be adopted in this monograph.

where $i = \sqrt{-1}$. The real part of $V^*(t)$ equals $V(t)$ as defined by equation (1.13), and we assume that all relevant physical laws hold for the complex function $V^*(t)$ as for its real part. We furthermore assume a similarly defined complex quantity $Q^*(t)$ for the charge and put

$$Q^*(\omega, t) = C^*(\omega)V^*(\omega, t). \tag{1.15}$$

$C^*(\omega)$ denotes a quantity which is time independent and will here be called a complex capacitance, since this term illustrates the dielectric analogy. When

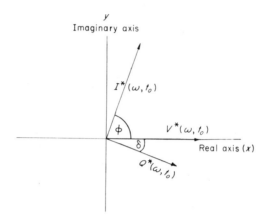

Fig. 2. Charge, current and voltage represented in the complex plane, at an instant of time $t = t_0$.

the complex quantities are substituted in equation (1.9) it is found that for

$$C^*(\omega) = \frac{C_s}{1 + i\omega\tau} \tag{1.16}$$

equation (1.15) is a solution of the differential equation (1.9). It is not the most general solution for a periodic voltage: this is quoted as equation (2.17) in Chapter 2. Equations (1.14)–(1.16) presuppose that the voltage has been periodic for an infinite time.

The significance of the complex capacitance is most clearly seen if complex voltages, charges, etc., are represented in the complex plane, where a quantity $Z^* = x + iy$ is represented by two orthogonal coordinates x and y. In this plane the voltage $V^*(t)$ is a vector which rotates anti-clockwise with a period ω (equation 1.14). When $C^*(\omega)$ is separated into its real and imaginary part we may write for the charge

$$Q^*(\omega, t) = C_s V_0 \left(\frac{1}{1 + \omega^2\tau^2} - i \frac{\omega\tau}{1 + \omega^2\tau^2} \right) e^{i\omega t} \tag{1.17}$$

1. RELAXATION IN ELECTRICAL CIRCUITS

and it is seen that the charge is also represented by a rotating vector; as is also the current, given by

$$I^*(\omega, t) = \frac{dQ^*(\omega, t)}{dt} = i\omega C^*(\omega) V^*(\omega, t). \tag{1.18}$$

Figure 2 shows a stationary picture of the synchronously rotating vectors $V^*(\omega, t)$, $Q^*(\omega, t)$ and $I^*(\omega, t)$, taken at a moment when $t = 2\pi n/\omega$ and the voltage is at its peak. It may be seen that when $V^*(\omega, t)$ is real $Q^*(\omega, t)$ is partly imaginary, and that it lags behind the voltage. The angle δ between the charge and voltage vectors follows from equation (1.17) by

$$\tan \delta = \omega \tau. \tag{1.19}$$

Comparison of equations (1.17) and (1.18) shows that

$$\phi + \delta = \frac{\pi}{2}. \tag{1.20}$$

Differentiation corresponds to a multiplication by $i\omega$ and a rotation by a right angle in the complex plane.

The use of a complex capacitance is not conventional in circuit theory, where the generalization to complex variables is applied not to equation (1.1) but to Ohm's law, so that

$$V^*(\omega, t) = Z^*(\omega) I^*(\omega, t) \tag{1.21}$$

where $Z^*(\omega)$ is the complex impedance. Comparison of equations (1.18) and (1.21) gives

$$Z^*(\omega) = \frac{1}{i\omega C^*(\omega)} \tag{1.22}$$

and in the following we shall mostly operate with impedances, as customary. The impedance may conveniently be written in complex polar coordinates as

$$Z^*(\omega) = Z(\omega) \, e^{i\phi} \tag{1.23}$$

where $Z(\omega)$ is real, and often called simply the impedance of the circuit. For the circuit in Fig. 1

$$Z(\omega) = \frac{1}{\omega C_s} \sqrt{(1 + \omega^2 \tau^2)} \tag{1.24}$$

according to equations (1.16) and (1.22).

At this stage it is instructive to express the charge in polar complex coordinates, when

$$Q^*(\omega, t) = \frac{V_0}{\omega Z(\omega)} e^{i(\omega t - \delta)} \tag{1.25}$$

according to equations (1.14), (1.15), (1.20), (1.22) and (1.23). This expression illustrates the meaning of τ. For small values of tan δ equation (1.19) implies that $\delta = \omega\tau$ in fair approximation, so that τ measures a time lag between the charge and the voltage.

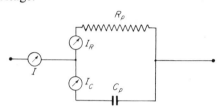

FIG. 3. Parallel combination of a capacitor and a resistor.

The time lag of the charge is connected with a dissipation of energy. The change, with time, of the energy present in the circuit is at any time given by the product of the (real) voltage and the (real) current

$$W(t) = V(t) \cdot I(t). \tag{1.26}$$

The value of $W(t)$ may be positive or negative. Its significance will be discussed in Chapter 3. Here it is important that in the average over a full cycle the rate of energy dissipation is given by

$$W = \tfrac{1}{2} V_0^2 \frac{\sin \delta}{Z(\omega)} \tag{1.27}$$

an expression which applies generally to linear passive networks and sinusoidal currents. For small angles δ the energy dissipation or power loss is directly proportional to δ. Hence δ is commonly called the loss angle and tan δ the loss tangent.

A capacitor in series with a resistor is analogous to other systems in physics or chemistry. In particular it is analogous to a water reservoir of given volume which is filled through a channel in which the stream is subject to friction. A much used analogy in mechanics consists of a spring in parallel with a dashpot. Here V corresponds to a stress (applied for instance by a hanging weight) while Q is the strain, measured by the length of the spring. The strain follows the stress with a delay and the same differential equation holds as for the circuit in Fig. 1.

A second very simple electrical circuit important for relaxation phenomena consists of a capacitor in parallel with a resistor (Fig. 3). In this case it is

1. RELAXATION IN ELECTRICAL CIRCUITS

convenient to derive a differential equation in terms of currents, namely

$$I = I_R + I_C = \frac{V}{R_p} + C_p \frac{dV}{dt} \quad (1.28)$$

where V and I are measured as indicated in the figure. For sinusoidal currents the solution of the differential equation (1.28) by the method outlined above leads to an impedance

$$Z^*(\omega) = \frac{R_p}{1 + i\omega C_p R_p} \quad (1.29)$$

and a loss tangent

$$\tan \delta = \frac{1}{\omega C_p R_p} \quad (1.30)$$

which characterize the relationship between charge and voltage.

The parallel and series combinations of capacitance and resistance are equivalent to each other with regard to a sinusoidal applied voltage, in the sense that a measurement at a constant frequency cannot distinguish between them. This means that if $Z^*(\omega)$ for a black box with two terminals has been measured at a single frequency the contents of the box may be equally well represented by assuming a hypothetical series circuit as a hypothetical parallel circuit. The relationship between C_s and R_s and its equivalent C_p and R_p for a given $Z^*(\omega)$ is derived in Appendix II, in conjunction with formulae for the impedances of circuits.

The parallel and series combinations however are not equivalent with regard to a step-function in voltage. For the parallel combination a step-function in voltage with $V = 0$ for $t \leq 0$ and $V = V_0$ for $t > 0$ implies that an infinite current flows for an infinitely short time through the capacitor while a constant current flows through the resistor for $t > 0$. This means that a step-function may distinguish between the series and parallel combination as discussed in Section A of Chapter 6. When a step-function in voltage is applied to the parallel combination of capacitance and resistance no finite relaxation time can be defined and equation (1.28) is not a relevant description of the process in question.

In general, a relaxation time τ is an attribute of a differential equation

$$\tau \frac{dx}{dt} + x = y \quad (1.31)$$

and such an equation is a short description of a process. In some cases differential equations with different values of τ may be set up for processes in the same circuit. A simple example follows.

Figure 4 shows a combination of two capacitors and a resistor, a circuit which is an analogue of the most important theoretical model of a dielectric material. The charge held between the terminals is the sum of the charge on the two capacitors

$$Q = Q_1 + Q_2 \tag{1.32}$$

and two separate differential equations may be defined for Q_1 and Q_2 as a function of the voltage V at the terminals, namely

$$Q_1 = C_1 V \tag{1.33}$$

$$R_2 \frac{dQ_2}{dt} + Q_2 = C_2 V. \tag{1.34}$$

For alternating currents the complex capacitance may be derived for the

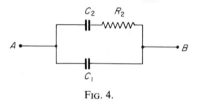

Fig. 4.

relationship of the charge Q and the voltage between the terminals. The real and imaginary parts of the complex capacitance $C^*(\omega)$ are given by

$$C'(\omega) = C_1 + \frac{C_2}{1 + \omega^2 \tau^2}$$
$$C''(\omega) = \frac{C_2 \omega \tau}{1 + \omega^2 \tau^2} \tag{1.35}$$

where

$$\tau = C_2 R_2 \tag{1.36}$$

is the relaxation time of the branch containing C_2 and R_2 in series. The same two differential equations apply to a step-function in voltage, where the infinitely rapid charging of C_1 and the delayed charging of C_2 via R_2 are the relevant processes.

It is possible to define another process in the circuit of Fig. 4 as follows: at a time $t = 0$ a voltage V_0 is applied for an infinitely short time, leaving a charge Q_0 on the plates of capacitor C_1, while capacitor C_2 is not charged. Then the external leads are taken away and the system is left to itself. Now the voltage across the terminals A and B gradually decreases as the charge distributes itself between the two capacitors.

For the process in question a differential equation for the charge Q_1 follows from

$$Q_1 + Q_2 = Q_0 \tag{1.37}$$

$$\frac{dQ_1}{dt} = I \tag{1.38}$$

$$IR_2 = V_1 - V_2 = \frac{Q_1}{C_1} - \frac{Q_2}{C_2} \tag{1.39}$$

which gives

$$R_2 \frac{dQ_1}{dt} + Q_1 \frac{C_1 + C_2}{C_1 C_2} = \frac{Q_0}{C_2}. \tag{1.40}$$

This is a differential equation of type (1.31) but it has a relaxation time

$$\tau = \frac{C_1 C_2 R_2}{C_1 + C_2} \tag{1.41}$$

which is different from the relaxation time given by (1.36). For this process both capacitances appear in series with the resistance R_2. The time-dependent behaviour of this circuit has been considered in a more general way by Gross and Pelzer[3] who treated the infinitely rapid charging of C_1 by means of Dirac δ functions and were concerned with the mathematical aspects. The example of the two relaxation times for the simple circuit shown in Fig. 4 demonstrates that a relaxation time is not a property of a circuit as such, but of a process in a circuit.

Equations (1.33) and (1.34) are effectively two differential equations for two branches of a network, where branch 1 has $\tau = 0$. In general it is not possible to write down a single differential equation of type (1.31) with a single value of τ for any network with two terminals. However, it is always possible to derive an impedance. The significance of this fact is illustrated by considering the slightly more complicated network in Fig. 5(a), for which

$$Z^*(\omega) = \frac{(1 - i\omega(C_1 + C_2))(R_2 + R_3)}{1 + \omega^2 C_2^2 R_2^2} \tag{1.42}$$

This expression implies a quadratic term in ω which does not cancel out on division. Now a term ω^2 arises by double differentiation (see equation 1.18), and the differential equation solved on the basis of the complex impedance is of the second order; that is, it contains a term in $d^2 I/dt^2$. This mathematical point was made already by Maxwell, in a more elegant way, in a context quoted in Chapter 14. However, higher order differential equations for networks containing only capacitances and resistances are not of practical interest. It is always possible[2] to represent a given network such as that shown in Fig. 5(a) by an equivalent one, such as that shown in Fig. 5(b), where each branch can be described by a first-order differential equation, so

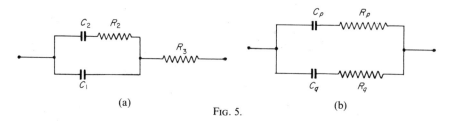

FIG. 5.

that a differential equation of the nth order is replaced by n differential equations of the first order, of type (1.31), each with a constant relaxation time τ. Circuits which are defined as equivalent in this sense behave in the same way with regard to any voltage $V(t)$ applied to their terminals[2].

B. THE DIFFERENTIAL EQUATION FOR THE SIMPLEST RESONANT CIRCUIT

For a network containing inductances L as well as capacitances and resistances, a linear second-order differential equation is the simplest differential equation capable of describing the processes in question. An *RCL* circuit is the appropriate analogue for a dielectric material at optical frequencies but is only very rarely relevant at frequencies used in electrical engineering or electronics. However, the role of inductance is of sufficient importance to be briefly discussed here.

The differential equation for a circuit containing resistance, capacitance and inductance all in series may be be derived as for the series combination of capacitance and resistance. Addition of the voltages in series gives

$$L\frac{dI}{dt} + RI + \frac{1}{C}\int I\,dt = V(t) \qquad (1.43)$$

which may be arranged to read

$$\omega_0^2 \frac{d^2 I}{dt^2} + \tau_0 \frac{dI}{dt} + I = C\frac{dV}{dt} \qquad (1.44)$$

where

$$\omega_0 = \frac{1}{\sqrt{(LC)}} \qquad (1.45)$$

$$\tau_0 = RC \qquad (1.46)$$

are constants. The solution of a differential equation of this type for sinusoidal currents will be discussed in Chapter 9 in connection with resonance effects. Here it will be instructive to discuss the solution of equation (1.44) for a step-function. For $V = V_0$ for $t < 0$ and $V = 0$ for $t > 0$ (that is, a sudden discharge), the solution for $Q(t)$ is given by

$$Q(t) = A_1 e^{k_1 t} + A_2 e^{k_2 t} \tag{1.47}$$

where A_1 and A_2 are constants and k_1 and k_2 are solutions of the quadratic equation

$$k_{1,2} = -\frac{R}{2L} \pm \sqrt{\left(\frac{R^2}{4L^2} - \frac{1}{LC}\right)}. \tag{1.48}$$

This means that after the switching off of the voltage V_0 the charge on the capacitor may decay in one of two different ways.

If

$$\frac{R}{2L} < \frac{1}{\sqrt{(LC)}} \tag{1.49}$$

the coefficients k_1 and k_2 are complex, and the discharge is oscillatory. If $R = 0$ the oscillations persist indefinitely, and have a period ω_0 (equation 1.45). For finite resistance the oscillations are damped so that

$$Q(t) = CV_0 e^{-\alpha t} \cos(\omega_d t) \tag{1.50}$$

where

$$\alpha = \frac{R}{2L} \tag{1.51}$$

is the damping factor and

$$\omega_d = \sqrt{(\omega_0^2 - \alpha^2)} \tag{1.52}$$

is the frequency of the damped oscillations. For

$$\frac{R}{2L} \geq \frac{1}{\sqrt{(LC)}} \tag{1.53}$$

the charge decays aperiodically, but with two relaxation times. In the case where the two coefficients k_1 and k_2 are equal the discharge is called critically damped.

Here

$$Q(t) = A e^{-\alpha t} \tag{1.54}$$

where

$$\alpha = \frac{R}{2L} = \frac{2}{RC}. \tag{1.55}$$

Expression (1.54) is identical with that for the series combination of capacitance and resistance (equation 1.12) except for the magnitude of the time constant. For the critically damped case of the series RCL circuit

$$\tau = \frac{RC}{2} \tag{1.56}$$

that is, only half as long as it would be for the case of the series RC circuit (equation 1.10).

Relaxation in electrical networks is only one of the several macroscopic phenomena governed by a differential equation of the type (1.31) or (1.44). The mathematical treatment of all these phenomena is obviously identical, and solutions can be translated from one physical system to another[4]. The present monograph is not concerned with all of them, but some types of mechanical relaxation give information that complements the electrical data on microscopic systems, and it seems worth while to give here the appropriate translation of a simple macroscopic case.

Assume a gas or liquid or amorphous solid of given volume v, subject to a pressure p, the change of volume due to the pressure being

$$x = v(p) - v(p_0) \tag{1.57}$$

where p_0 is a standard pressure. Then x obeys an equation of type (1.31) or (1.44), where the stimulus y is proportional to $p(t) - p_0$. The term L is represented by a quantity M of dimension g cm^{-4}, called "inertance", C by an acoustical capacitance C_A of dimension g^{-1} cm^4 sec^2, and R by an acoustical resistance R_A of dimension g cm^{-4} sec^{-1}. The inertia in this case, as in all mechanical cases, is due to a mass.

In a gas the acoustical system is essentially periodic dominated by the inertia term. However, in liquids and particularly in soft solids the damping may be the dominant influence. In those cases the material and its properties may be described without reference to the inertia term. Here the mathematics is quite analogous to the case of an RC network. The relevant properties are usually measured by the application of periodic pressure waves, using the techniques of ultrasonics.

REFERENCES

1. M. Abraham and R. Becker (1932). "The Classical Theory of Electricity and Magnetism." Blackie & Son, London.
2. E. A. Guillemin (1953). "Introductory Circuit Theory." Chapman and Hall, London.
3. B. Gross and H. Pelzer (1951). *Proc. R. Soc. Ass.*, **210**, 434.
4. H. F. Olson (1958). "Dynamic Analogies." D. van Nostrand, London.

CHAPTER 2

Relaxation in Dielectrics with a Single Relaxation Time

A. MACROSCOPIC DERIVATION OF THE DEBYE EQUATIONS

The present chapter is concerned chiefly with the response of matter on a molecular scale to electric fields. Molecular phenomena will be considered in terms of simple models. However, before embarking on the main theme we shall recapitulate the elementary electrostatics involved and link these up with the concepts used in circuitry.

In Chapter 1 we considered capacitors whose capacitance was completely time independent; that is, a charge Q depended on voltage simply as

$$Q = \varepsilon C_0 V \qquad (2.1)$$

where C_0 was a geometrical factor and ε a constant with respect to time. In fact a capacitor with a material dielectric never behaves entirely in this way. For changing voltages the charge responds with a delay, so that one might be tempted to define a real function $\varepsilon = \varepsilon(t)$ for all dielectrics other than vacuum, even though this is mathematically not expedient. The delayed response of matter is due to the procedure by which it causes an increase of capacitance. A dielectric constant $\varepsilon > 1$ implies that displacements occur in the dielectric, and these always need some time. In consequence, a capacitor containing a dielectric behaves by itself like an RC circuit, and sometimes like an RCL circuit. Consideration of the significance of the dielectric constant shows the analogy to be very close, in that the same differential equations apply to the real capacitor as to a fictitious circuit which may be treated as its analogue. This means that the dielectric constant appears as a complex quantity analogous to an impedance.

In order to arrive at the mathematical formulation of the dielectric constant it is convenient to work in terms of fields and charge densities[1]. For a plate capacitor with parallel plates (edge effects being neglected) we define the electric field E as

$$E = \frac{V}{d} \qquad (2.2)$$

where d is the distance between the plates and V the potential difference between them.

The charge density is defined as

$$\sigma = \frac{Q}{A} \tag{2.3}$$

where A is the area. Elementary electrostatics show that field and charge density are connected by

$$E = 4\pi\sigma_0 \tag{2.4}$$

where σ_0 is the charge density for the capacitor in vacuo. The factor 4π is due to the use of the c.g.s. system (see Appendices I and II).

When a dielectric is introduced into the capacitor while the external voltage is kept constant, more charge flows into the capacitor whose storage capacity is increased by the presence of the dielectric. We shall assume that infinite time is allowed for the introduction of the dielectric and of the extra charge and that finally the charge density is

$$\sigma = \varepsilon_s \sigma_0 \tag{2.5}$$

where ε_s is the static dielectric constant, representing equilibrium after infinite time.

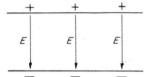
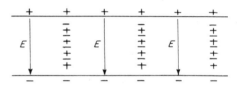

FIG. 1. (a) A capacitor in vacuum. The charges on the plates cause an electrical field E in the space between the plates; (b) A capacitor filled with a dielectric material: a part of the charges on the plates is compensated by a displacement of charges within the dielectric and does not cause a field E in the space between the plates.

The increased capacitance is due to the polarization of the dielectric in the sense that positive and negative charges within it are displaced slightly from their normal positions. Greater or lesser displacements of this kind occur in every material, since matter is built up from charged units.

The meaning of polarization is illustrated in Fig. 1, which shows that the polarization of matter in a capacitor acts in the same way as a field, in that it holds charges on the electrodes. The polarization and the external field combine to hold the total charge Q.

In dielectric theory it is usual to introduce a field quantity D

$$D = 4\pi\sigma \tag{2.6}$$

where σ is the total charge density which is held on the plates by the combined action of external field and polarization.

2. RELAXATION IN DIELECTRICS WITH A SINGLE RELAXATION TIME

The polarization P is defined by

$$\sigma_0 + P = \sigma \qquad (2.7)$$

as the charge density held on the plates by the internal displacement of charges. In view of equations (2.4), (2.5) and (2.6), equation (2.7) transforms to

$$E + 4\pi P = D \qquad (2.8)$$

while equation (2.5) leads to

$$D = \varepsilon_s E. \qquad (2.9)$$

Combining (2.8) and (2.9) we have a definition for the static dielectric constant in terms of the polarization which is here the static value P_s

$$\varepsilon_s - 1 = \frac{4\pi P_s}{E}. \qquad (2.10)$$

The polarization P_s implies a dipole moment of the dielectric in the whole of the capacitor of

$$M_s = P_s A d \qquad (2.11)$$

so that P_s can also be defined as a dipole moment/unit volume. A dipole moment signifies the product of a charge and a distance, and is a concept important for the interpretation of dielectric properties on a microscopic scale.

To continue with macroscopic properties, equation (2.10) gives a relationship between P_s and E in equilibrium. If E changes with time, P at a given moment will generally differ from $P_s(E)$. A trend towards equilibrium implies that P will approach P_s and an assumption about the rate of change of P as a function of time will give a differential equation for $P(t)$. The simplest assumption is to take the speed of approach to equilibrium proportional to the distance from equilibrium, so that

$$\tau \frac{dP(t)}{dt} = P_s - P(t) \qquad (2.12)$$

with τ constant. Later considerations will show this assumption to be very plausible.

Before integrating equation (2.12) we shall introduce a modification to (2.10). Several mechanisms of polarization operate normally in a dielectric, and one of these is the electronic polarization, which involves resonance and is responsible for the optical refractive index. This polarization responds to a changing field in negligible time at all frequencies used for electrical measurements, and hence may be considered as not entering equation (2.12). To separate it out we write instead of equation (2.10)

$$\varepsilon_s - 1 = \frac{4\pi}{E}(P_D + P_\infty) \qquad (2.13)$$

where P_D now refers only to the static "dipolar" polarization to be discussed later, while P_∞ is the optical polarization, and defines an optical dielectric constant ε_∞ by

$$\varepsilon_\infty - 1 = \frac{4\pi P_\infty}{E} = n^2 - 1 \qquad (2.14)$$

where n is the refractive index. There is some ambiguity here with regard to the frequency at which n should be measured, since frequencies generated by microwave oscillators extend up to the order of 10^{11} sec^{-1}, where resonance phenomena begin to appear (see Chapter 9), and since the optical refractive index is measured for frequencies of 10^{14}–10^{15} sec^{-1}. However, it is usually satisfactory to consider ε_∞ as the square of the refractive index measured at frequencies around 10^{10}–10^{12} sec^{-1}.

When an electric field $E(t)$ is applied to a dielectric, $P_\infty(E)$ responds instantly, while P_D responds according to an equation corresponding to (2.12)

$$\tau \frac{dP_D}{dt} + P_D(t) = (\varepsilon_s - \varepsilon_\infty)\frac{E(t)}{4\pi} \qquad (2.15)$$

where the right-hand side of the equation is the equilibrium value of P_D corresponding to $E(t)$, i.e. that value of P_D which would be reached if the instantaneous value $E(t)$ of the field were applied for an infinite time.

Equation (2.15) is analogous to equation (1.7) for the charge on a capacitor in series with a resistor. The general solution of equation (2.15) (or 1.7) for a periodic field

$$E^*(t) = E_0 e^{i\omega t} \qquad (2.16)$$

needs to take into account the fact that at $t = 0$, the dielectric may have been polarized as a consequence of its previous history. The general solution of equation (2.15) may be expressed in terms of a complex quantity P_D^* for the dipolar polarization

$$P_D^*(t) = K e^{-\frac{t}{\tau}} + \frac{1}{4\pi}\frac{\varepsilon_s - \varepsilon_\infty}{1 + i\omega\tau} E_0 e^{i\omega t} \qquad (2.17)$$

where K characterizes the initial polarization. It may be seen that the first term decays with time, and for alternating current measurements it can in

2. RELAXATION IN DIELECTRICS WITH A SINGLE RELAXATION TIME

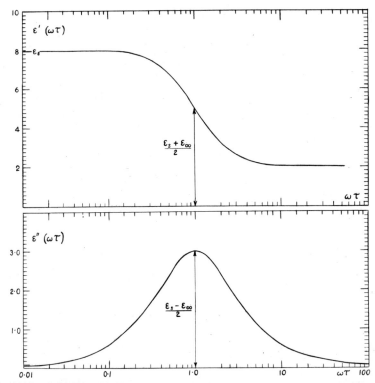

FIG. 2. (a) The real part of ε^* as a function of $\log \omega\tau$ according to the Debye equations (2.21) and (2.22) for a dielectric with $\varepsilon_s = 8, \varepsilon_\infty = 2$; (b) The imaginary part of ε^* for the same dielectric.

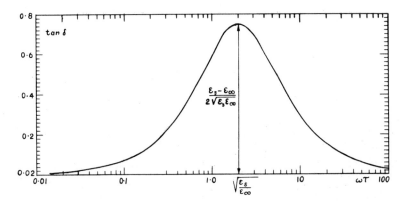

FIG. 3. The loss tangent for the dielectric illustrated in Fig. 2(a) and (b).

practice be neglected in comparison with the second term. When the first term is neglected, we may define a dielectric constant

$$\varepsilon^*(\omega) - \varepsilon_\infty = 4\pi \frac{P_D^*(\omega, t)}{E^*(\omega, t)} \quad (2.18)$$

where asterisks denote complex quantities. This treatment is analogous to the treatment of the differential equation (1.9) by way of a complex capacitance. Inspection of equation (2.17) shows that for $K = 0$

$$\varepsilon^*(\omega) - \varepsilon_\infty = \frac{\varepsilon_s - \varepsilon_\infty}{1 + i\omega\tau} \quad (2.19)$$

where $\varepsilon^*(\omega)$ may be separated into its real and imaginary part.

Since

$$\varepsilon^*(\omega) = \varepsilon'(\omega) - i\varepsilon''(\omega) \quad (2.20)$$

we have

$$\varepsilon'(\omega) = \varepsilon_\infty + \frac{\varepsilon_s - \varepsilon_\infty}{1 + \omega^2\tau^2} \quad (2.21)$$

$$\varepsilon''(\omega) = (\varepsilon_s - \varepsilon_\infty)\frac{\omega\tau}{1 + \omega^2\tau^2}. \quad (2.22)$$

Equations (2.21) and (2.22) are usually called Debye equations since they were derived by Debye on a molecular basis, discussed later in this chapter.

The functions represented by equations (2.21) and (2.22) are illustrated in Fig. 2, plotted against $\log_{10}\omega$, for the specific values $\varepsilon_s = 8$ and $\varepsilon_\infty = 2$.

Experimentally one measures usually the quantities $\varepsilon'(\omega)$ and the loss tangent

$$\tan \delta = \frac{\varepsilon''(\omega)}{\varepsilon'(\omega)} \quad (2.23)$$

and it is convenient to know the Debye equations in terms of the loss tangent, when

$$\tan \delta = \frac{(\varepsilon_s - \varepsilon_\infty)\omega\tau}{\varepsilon_s + \varepsilon_\infty\omega^2\tau^2}. \quad (2.24)$$

Figure 3 gives $\tan \delta$ as a function of frequency for the same values as those used in Fig. 2. It may be seen that the position of the maximum of $\tan \delta$ is not identical with the position of the maximum of $\varepsilon''(\omega)$.

Equations (2.21) and (2.22) for the dielectric constant are analogous to equations (1.34) and (1.35) for the complex capacitance connecting charge and voltage in the circuit shown in Fig. 4. In fact this circuit is, in its response to applied fields, completely analogous to a dielectric obeying the Debye

2. RELAXATION IN DIELECTRICS WITH A SINGLE RELAXATION TIME

equations as long as its parameters are chosen as follows: $C_1 = \varepsilon_\infty C_0$, $C_2 = (\varepsilon_s - \varepsilon_\infty)C_0$ and R given by $C_2 R = \tau$ where C_0 is the empty capacitance of the capacitor containing the dielectric.

A dielectric with a single relaxation time may also be represented by a parallel combination of capacitance and resistance, but R_p and C_p depend on ω as shown in Appendix II. The representation of a dielectric by a parallel combination of capacitance and resistance is often practically convenient. We shall give the relevant formulae in practical c.g.s. units, when C_0 is the capacitance of a plate capacitor of area 1 cm^2 and 1 cm distance between plates, while $1/R$ is now the specific conductivity γ and measured in Ω^{-1} cm^{-1}. The relationship is

$$\gamma(\omega) = \frac{\omega \varepsilon''(\omega)}{3 \cdot 6\pi} 10^{-12}. \tag{2.25}$$

Although $\varepsilon''(\omega)$ of a Debye dielectric goes to zero for $\omega \to \infty$ the equivalent parallel conductivity increases with frequency, rising to

$$\gamma_\infty = \frac{\varepsilon_s - \varepsilon_\infty}{3 \cdot 6\pi\tau} 10^{-12}. \tag{2.26}$$

This signifies that for infinite frequency all the field energy is dissipated as heat (see Section C of Chapter 3).

B. MOLECULAR MODELS WITH A SINGLE RELAXATION TIME

The differential equation (2.15) which is the basis for the Debye equation may be derived from models of the microscopic constitution of matter. Not all possible models lead to this equation, but it is by far the most useful basis for the treatment of dielectric relaxation, as well as the simplest one. We shall in the following treat in some detail two of the many possible models which lead to the Debye equations. Other models are considered later in connection with special types of dielectrics.

The main subject of the theory of the static dielectric constant[3] is the calculation of the dipole moment M_s of a volume of dielectric (see equation 2.11) for a given field E from the configuration of positive and negative charges on a molecular or atomic, that is microscopic, scale. In this context it is usual to speak of dipoles in the sense of two electric charges of magnitude $+e$ and $-e$ a small distance l apart. A dipole in this sense is a vector and can respond to a field by altering the length l or the direction of l. The length may be $l = 0$ for zero field, when the molecule is said to have no permanent moment. This representation implies that individual charges cannot move over large distances, i.e. that there is no d.c. conductivity if a material is made up of dipoles alone, and does not contain free charge carriers. It is therefore a suitable representation of a dielectric. However, the concept of a permanent dipole with $l(E = 0) > 0$ is only a mathematical fiction in some dielectrics, even though it applies very well to others. Some chemical compounds contain molecular grouping of well-defined permanent dipole moment, for instance the $C = 0$ or keto group in organic chemistry where the positive charges are centred near to carbon and the negative ones near to oxygen. Many such moments have been tabulated, but in many materials permanent dipoles cannot be easily identified in terms of definite molecular configurations.

The task of dielectric theory is difficult not so much because permanent dipoles cannot always be identified, but mainly because they influence one another mutually; a dipole is not only subject to the influence of a field but also has a field of its own. The mutualness of the influence of dipoles, permanent or otherwise, on one another makes the response of the assembly a cooperative phenomenon and causes it to depend on the size and shape of the assembly. The effective field acting on a dipole is in general not the externally applied field E but is augmented by a contribution caused by cooperation. The local field acting on the dipole will be called E_i, where $E_i \geqslant E$. We shall at first consider cases where $E_i = E$, that is negligible electrostatic interaction.

The first model to be considered is probably the simplest and most useful of all and it has been treated by Fröhlich[3]. We shall call it the bistable model. For the purposes of this model it is assumed that a particle of charge e may be in one or other of two sites, 1 and 2, located a distance b apart. These sites are defined as minima of the potential energy as shown in Fig. 4. An electric field acting on the system causes a difference in the potential energy of the two sites, and the figure is drawn with solid lines for the condition without field, and with dotted lines for the presence of a field.

The potential difference due to the field E is

$$\phi_1 - \phi_2 = e(bE) = ebE \cos \theta \qquad (2.27)$$

where θ is the angle between the line 1–2 and the direction of the field. This model is equivalent to a dipole in that a movement of the charge from 1 to 2 or *vice versa* is equivalent to a turn by 180° of angle of a dipole of moment

$$\mu = \tfrac{1}{2}eb \qquad (2.28)$$

which one might imagine to be hinged about the centre point between 1 and 2.

We assume a number N of bistable dipoles/unit volume representing a small density, so that the field due to dipolar interaction can be neglected. We also assume $\cos \theta = 1$ for all dipoles, and equal potential energy for the

2. RELAXATION IN DIELECTRICS WITH A SINGLE RELAXATION TIME

sites 1 and 2 in the absence of an electric field. We shall quote the results for two more general cases.

The model described would have no dynamic properties if it were on a macroscopic scale because the charged particles in question would not have the energy to jump the "potential hill" between the two "potential wells" 1 and 2. However, in a microscopic assembly the N bistable dipoles must be imagined as being located in a heat reservoir which consists of spontaneously

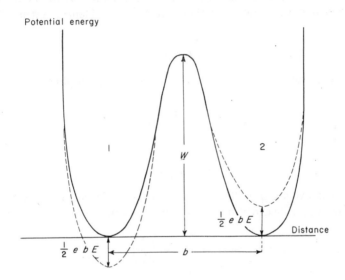

FIG. 4. The bistable model. The potential energy as a function of distance has two minima, "potential wells", whose depth is modified by an applied field E. The left-hand well is denoted by 1, the right-hand well by 2. The two wells contain one charged particle which may occupy either well.

active particles which exchange energy with each other and the dipoles. Hence the directions of the dipoles fluctuate. A charge situated in a well 1 occasionally acquires an energy sufficient to lift it over the potential hill, and the charge drops into the well 2 associated with it. On arrival in 2 the energy of the charge is returned to the heat reservoir, and the charge then stays in 2 until such time as it acquires enough energy from the reservoir to return over the hill to 1. The probability of a jump by a charge in a double potential well can be derived from statistical thermodynamics, as discussed in Chapter 4. Here it is sufficient to state that the number of dipoles jumping/unit time from 1 to 2 is given in terms of the difference of potential energy between the two wells as

$$w_{12} = A\, e^{-\frac{W - \mu E}{kT}} \qquad (2.29)$$

where T is the absolute temperature, $k = 1.37 \times 10^{-16}$ erg/°C is the Boltzmann constant, and A a factor which may or may not depend on temperature (see Section E of Chapter 4). The minus sign holds if well 2 is higher than well 1, as shown in the figure where the field points towards the left. Equation (2.29) presupposes that $W \gg kT$; the energy W is usually described as an activation energy.

At this stage the magnitude of μE becomes important. For dipoles in normal dielectrics μ is of the order 10^{-18} e.s.u., while E for fields below breakdown strength is always smaller than 10^5 e.s.u. Hence, in general,

$$\frac{\mu E}{kT} \ll 1 \tag{2.30}$$

and the expression for the frequency of jumps can be simplified to

$$w_{12} = w\left(1 - \frac{\mu E}{kT}\right) \tag{2.31}$$

where

$$w = A e^{-\frac{W}{kT}} \tag{2.32}$$

is the frequency of jumps in the absence of an applied field. The frequency of jumps in the opposite direction, from 2 to 1, is in the sketched case given by

$$w_{21} = w\left(1 + \frac{\mu E}{kT}\right). \tag{2.33}$$

The average population of charges in wells 1 and 2 of the N bistable dipoles will not change with time if the number of charges jumping/unit time, from left to right, equals that jumping from right to left; i.e., if

$$N_1 w_{12} = N_2 w_{21} \tag{2.34}$$

where N_1 is the number of occupied wells 1 and N_2 that of occupied wells 2. Since the total number of occupied wells/unit volume, i.e. the total number of bistable dipoles, is constant, we have

$$N_1 + N_2 = N. \tag{2.35}$$

Equations (2.34) and (2.35) permit the calculation of N_1 and N_2 in equilibrium, and hence the polarization in equilibrium. In general, the polarization/unit volume is given by that number of dipoles in one direction which is not compensated by dipoles in the opposite direction, namely

$$P = (N_1 - N_2)\mu. \tag{2.36}$$

2. RELAXATION IN DIELECTRICS WITH A SINGLE RELAXATION TIME

In equilibrium $N_1 - N_2$ and hence $P = P_s$ can be calculated from equations (2.34) and (2.35), using equations (2.31) and (2.33). We have

$$N_1 w\left(1 - \frac{\mu E}{kT}\right) = N_2 w\left(1 + \frac{\mu E}{kT}\right) \tag{2.37}$$

$$N_1 - N_2 = (N_1 + N_2)\frac{\mu E}{kT}$$

and thus, in view of (2.35) and (2.36)

$$P_s = N \frac{\mu^2 E}{kT}. \tag{2.38}$$

Equation (2.38) can be generalized for the case where the field makes an angle θ with the direction of the dipoles and large dipoles where equation (2.30) is not valid. In that case

$$P_s = \mu \cos \theta \tanh \frac{\mu E \cos \theta}{kT}. \tag{2.39}$$

The latter equation implies saturation of P_s with increasing field strength, and thus a non-linear relationship between P and E. Equation (2.38) may also be generalized for the case where electrostatic interaction is not negligible[3], but this is not an easy task.

The time-dependent properties of the model follow from the fact that the change in the number of dipoles in 1 is equal to the outflow to 2 less the inflow from 2, thus

$$\frac{dN_1}{dt} = -N_1 w_{12} + N_2 w_{21} \tag{2.40}$$

while the constancy of $N_1 + N_2$ (equation 2.36) gives

$$\frac{dN_2}{dt} = -\frac{dN_1}{dt}$$

and hence

$$\frac{d(N_1 - N_2)}{dt} = 2\frac{dN_1}{dt}. \tag{2.41}$$

Using the expansions of the probabilities for small argument (equations 2.31 and 2.33) we have

$$\frac{1}{2}\frac{d(N_1 - N_2)}{dt} = -wN_1\left(1 - \frac{\mu E}{kT}\right) + wN_2\left(1 + \frac{\mu E}{kT}\right) \tag{2.42}$$

$$= -w(N_1 - N_2) + w(N_1 + N_2)\frac{\mu E}{kT} \tag{2.43}$$

a differential equation for the argument $N_1 - N_2$ and, according to equation (2.36), for the polarization $P(t)$. A simple transformation gives

$$\frac{1}{2w} \cdot \frac{dP}{dt} + P = \frac{N\mu^2 E}{kT}. \tag{2.44}$$

This is a relaxation equation, a special case of equation (2.15), with a relaxation time

$$\tau = \frac{1}{2w} = \frac{1}{2A} e^{-\frac{W}{kT}}. \tag{2.45}$$

Equation (2.44) may be generalized and written

$$\tau \frac{dP_D}{dt} + P_D = \alpha_D E \tag{2.46}$$

where α_D is the dipolar polarizability and P_D the dipolar polarization. For the bistable model the term "dipolar" has an exact meaning, but it is also used in a more vague sense.

The treatment given here for a symmetrical double potential well with $w_{12} = w_{21}$ in the absence of a field can easily be extended to the case where the two probabilities are unequal for reasons other than the applied field, so that without field

$$\frac{w_{12}}{w_{21}} = e^{\frac{B}{kT}} \tag{2.47}$$

B being a constant. For this case we have

$$\frac{1}{w_{12} + w_{21}} \cdot \frac{dP}{dt} + P = \frac{N\mu^2 E}{kT} \operatorname{sech}^2 \left(\frac{B}{2kT} \right) \tag{2.48}$$

which is still of the type (2.46).

The interest of equations (2.45) and (2.48) as compared with (2.15) is that we have now an interpretation of the relaxation time in physical terms. It is the reciprocal of the frequency of jumps across the potential hill. In other words the delay in response is due to the fact that we have to wait for a rare event.

The physical meaning of a relaxation time is illuminated from another angle in Debye's[4] classical treatment of dielectric relaxation in a dilute solution of dipolar molecules in a non-polar liquid. This is interesting because it gives a graphic picture of fluctuations in a three dimensional assembly and links up with the theory of Brownian movement. Fluctuations will be discussed in more detail in Chapters 4 and 8, and Debye's treatment, which is somewhat lengthy, will only be sketched here.

2. RELAXATION IN DIELECTRICS WITH A SINGLE RELAXATION TIME

Debye assumes a small concentration of dipolar molecules in a non-polar solvent, and we shall here, in addition, assume that electrostatic interaction between dipoles is negligibly small. It is assumed that no external forces are present which influences the positions of dipoles. In this situation and in the absence of an electrical field, both the positions and orientations of dipoles fluctuate at random as a consequence of thermal motion. An applied electrical field affects the randomness of orientation, but not of position.

The orientation of dipoles can be described by a density function f in terms of polar coordinates. Figure 5 shows a small solid angle $d\Omega$ on a unit sphere,

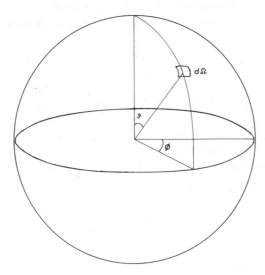

FIG. 5. The coordinates ϕ and ϑ of a solid angle $d\Omega$ sketched as it appears on the surface of a unit sphere.

delineated by coordinates ϕ and $\phi \pm d\phi$ and ϑ and $\vartheta \pm d\vartheta$. The angle is supposed to be large enough to contain enough dipoles to permit the definition of $f(\vartheta, \phi)$. In the absence of a field, the density/unit solid angle is independent of ϑ and ϕ, although dipoles drift in and out of $d\Omega$. Debye assumes a field in the direction $\vartheta = 0$. In the presence of this field more dipoles will point in the field direction than away from it. In equilibrium $f(\vartheta, \phi)$ will be a maximum for $\vartheta = 0$.

When a constant field is suddenly applied to a solution whose orientation is in equilibrium with a smaller field, the density of population of a given element $d\Omega$ changes with time. Now more dipoles drift into elements pointing in the field direction than drift out of them. The total change of polarization/unit volume of the solution as a whole may be derived by integration over all elements $d\Omega$.

Debye divides the drift in or out of a solid angle into two parts so that

$$\frac{\partial f}{\partial t} d\Omega \, dt = \Delta_1 + \Delta_2 \qquad (2.49)$$

where $f(\vartheta, \phi, t)$ is the density function and the drifts denote an excess of dipoles entering over those leaving. Now Δ_1 is defined as the drift produced by the applied field while Δ_2 is due to thermal (Brownian) motion (see Chapter 4). Δ_2 is calculated in terms of a quantity $\bar{\theta}^2/dt$, the mean square of the angle ϑ (in any direction) traversed/unit time by an average dipole in the course of thermal fluctuations. In Fig. 5 and angle ϑ would appear as a distance along the surface of the unit sphere. Debye shows that

$$\Delta_2 = \frac{d\Omega}{4} \bar{\theta}^2 \left(\tan \vartheta \cdot \frac{\partial f}{\partial \vartheta} + \frac{\partial^2 f}{\partial \vartheta^2} \right). \qquad (2.50)$$

This expression is zero if the density f of dipoles is independent of direction. However, if an applied field causes f to be dependent on the angle ϑ, thermal motion causes a drift which opposes the effects of the field.

Debye's calculation of Δ_1 is based on the fact that an electrical field causes a torque to act on a dipole but that the dipole is hindered in its rotation by a frictional force. Debye assumes that the dipole moves like a macroscopic body in a viscous liquid. This means that

$$-\mu E \sin \vartheta = \zeta \frac{d\vartheta}{dt} \qquad (2.51)$$

where ζ is a quantity characteristic of the friction, and constant.

The friction experienced by the dipole is an expression of the same thermal motion which causes its fluctuations of direction. The relationship between the friction constant and the magnitude of the fluctuations may be found from the condition that in equilibrium

$$\Delta_1 + \Delta_2 = 0. \qquad (2.52)$$

Debye shows that for the assumptions made (which include the assumption that the field does not change too fast compared with the fluctuations)

$$\frac{\bar{\theta}^2}{4dt} = \frac{kT}{\zeta}. \qquad (2.53)$$

Equation (2.53) is similar in character to the equation which relates the coefficient of diffusion D of particles in solution to their mean square displacement/unit time,

$$D = \frac{1}{2} \frac{\overline{x^2}}{dt}. \qquad (2.54)$$

2. RELAXATION IN DIELECTRICS WITH A SINGLE RELAXATION TIME

The quantity on the right-hand side may be determined experimentally by observing the Brownian movement of particles visible in an ultramicroscope.

Debye's treatment gives a relaxation equation for the function $f(\vartheta, t)$ of dipoles/unit solid angle. The relaxation time is given by

$$\tau = \frac{\zeta}{2kT}. \tag{2.55}$$

Since ζ is related to the fluctuation in angle τ might in principle be derived by means of observation of $\overline{\theta^2}$. However, this is not practicable for molecular dipoles. Debye chooses for the verification of his theory the expedient of relating ζ to the macroscopic shear viscosity. This approximation will be discussed in Chapter 11.

When the density function is integrated to give the polarization Debye's treatment leads to a differential equation of type (2.46) with τ given by equation (2.55) and

$$\alpha_D = \frac{N\mu^2}{3kT} \tag{2.56}$$

which is identical with α_D for the bistable model (equation 2.44) except for a factor $\frac{1}{3}$. This factor is due to the fact that Debye's dipoles point in all directions in space, while the bistable dipole has only two alternative positions. There are many possible dipole models with various alternative permitted positions and some of these will be discussed in the experimental chapters. They result in equations of type (2.46) which differ with regard to the constants τ and α_D.

C. AN IDEALIZED TREATMENT OF ELECTROSTATIC INTERACTION

So far we assumed that electrostatic interaction between dipoles was negligible. We shall now consider briefly how the electrostatic enhancement of the local field acting on a dipole affects the differential equation for relaxation. In general, electrostatic interaction causes considerable theoretical complications which will be discussed in Chapter 8. However, it may be treated relatively simply in a special case.

We consider an idealized dielectric, where the average field in the space between dipoles is given by

$$E_i = E + \frac{4\pi}{3} P \tag{2.57}$$

where E_i, E and P are vectors parallel to one another, so that their vector character need not be taken into account. The static macroscopic polarization is related to the applied field E and the dielectric constant by equation (2.13), while P_s may be expressed as a function of E_i according to

$$P_s = P_D + P_\infty = (\alpha_D + \alpha_\infty)E_i \qquad (2.58)$$

where α_D and α_∞ are the dipolar and optical polarizabilities. Equations (2.13), (2.57) and (2.58) give, after elimination of P_s,

$$(\varepsilon_s - 1)E = 4\pi(\alpha_D + \alpha_\infty)\left(E + \frac{\varepsilon_s - 1}{3}E\right)$$

or after rearrangement and division by E

$$\frac{\varepsilon_s - 1}{\varepsilon_s + 2} = \frac{4\pi}{3}(\alpha_D + \alpha_\infty). \qquad (2.59)$$

Equation (2.59) is usually called the Clausius–Mossotti equation. When equation (2.59) is rearranged to give ε_s explicitly it implies the so-called Clausius–Mossotti catastrophe which signifies that ε_s should tend to infinity as the quantity $1 - \frac{4\pi}{3}(\alpha_D + \alpha_\infty)$ tends to zero. This "catastrophe" is not observed in dielectrics other than ferroelectrics, and Chapter 8 shows that equation (2.59) does not hold for dipolar molecules in liquids. However, the Clausius–Mossotti equation holds moderately well[2] for cases where $\alpha_D = 0$, that is in conditions when the orientation of permanent dipoles does not contribute to the polarization. Hence the equation

$$\frac{\varepsilon_\infty - 1}{\varepsilon_\infty + 2} = \frac{4\pi}{3}\alpha_\infty \qquad (2.60)$$

can be used to express the average field acting within a non-polar dielectric in terms of ε_∞. Combination of equation (2.60) with equations (2.13) and (2.58) (with $\alpha_D = 0$) leads to

$$E_{i_\infty} = \frac{\varepsilon_\infty + 2}{3}E. \qquad (2.61)$$

This inner field is always in phase with E.

When a dielectric contains dipoles which respond to an applied field with a delay, the inner field E_i will also be delayed since it contains a contribution from the dipoles and this makes the derivation of a differential equation for relaxation difficult. The problem can still be solved for the idealized case in question since we may put

$$E_i = \left(E + \frac{4\pi}{3}P_D\right)\frac{\varepsilon_\infty + 2}{3} \qquad (2.62)$$

2. RELAXATION IN DIELECTRICS WITH A SINGLE RELAXATION TIME

where all the field vectors are parallel to one another but $P_D(t)$, $E_i(t)$ and $E(t)$ are no longer necessarily in phase. This equation means that the field acting on the dipoles is the instantly responding field E_{i_∞} enhanced still further by the lagging polarization $P_D(t)$.

On the basis of equation (2.62) for the field acting on the dipoles we can derive the dynamic response of the dipoles from

$$\tau_0 \frac{dP_D}{dt} + P_D = \alpha_D E_i$$

$$= \alpha_D \left(E + \frac{4\pi}{3} P_D \right) \cdot \frac{\varepsilon_\infty + 2}{3} \quad (2.63)$$

where τ_0 and α_D refer to an isolated dipole, that is a dipole responding to the external field, in the absence of any electrostatic interaction. Equation (2.63) transforms into

$$\tau_0 \frac{dP_D}{dt} + P_D \left(1 - \frac{4\pi \alpha_D}{3} \cdot \frac{\varepsilon_\infty + 2}{3} \right) = \alpha_D \frac{\varepsilon_\infty + 2}{3} E \quad (2.64)$$

which is a typical equation of type (2.46) but with values for polarizability and relaxation time different from the values for the isolated dipole. We may transform the constants in equation (2.64) by using equations (2.59) and (2.60). This gives an equation of type (2.46) with the constants

$$\tau = \tau_0 \frac{\varepsilon_s + 2}{\varepsilon_\infty + 2} \quad (2.65)$$

$$\alpha_D^e = \alpha_D \frac{\varepsilon_s + 2}{3}. \quad (2.66)$$

Equation (2.66) could be derived also by writing equation (2.59) in terms of (2.13) and a modified polarizability α_D^e.

The dielectric discussed responds to an applied external field as if electrostatic interaction were absent, but the polarizability/volume element and the relaxation time were modified according to equations (2.65) and (2.66), when compared with the intrinsic quantities τ_0 and α_D for an isolated dipole or other polarizable element. The quantities τ and α_D^e are called extrinsic as opposed to intrinsic.

The same result as that obtained above may also be obtained by way of a formal generalization of the Clausius–Mossotti equation for periodic fields when the intrinsic relaxation time τ_0 is ascribed to the dipolar polarizability α_D according to

$$\frac{\varepsilon^*(\omega) - 1}{\varepsilon^*(\omega) + 2} = \frac{4\pi}{3} \alpha_\infty + \frac{4\pi}{3} \alpha_D \frac{1}{1 + i\omega\tau_0}. \quad (2.67)$$

This equation transforms[2] into Debye equations (2.21) and (2.22) with a relaxation time τ given by equation (2.65).

As will be seen in Chapter 8 the idealized dielectric discussed here has no firm basis in theory or experiment. However, it may serve as an approximation for dielectrics whose structure is such that ferroelectricity is possible. Besides, equation (2.59) defines an upper limit for the effect which a number N of dipoles of moment μ can have on ε_s and $\tan \delta$, and thus a limit for the sensitivity of dielectric measurements (see Appendix III).

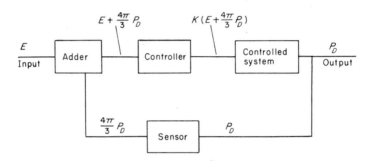

FIG. 6. A control system with positive feedback which is analogous to a dielectric obeying equation (2.64) with $\varepsilon_\infty = 1$.

The increase of polarizability and relaxation time due to electrostatic interaction is analogous to the behaviour of an electrical or mechanical system with positive feedback. Olson[5] considers systems with positive and negative feedback and their differential equations. The system characterized by equation (2.64) might be represented by the model in Fig. 6 where the "sensor", "adder", "controller" and "controlled system" would be "black boxes" with linear response. The quantities E, etc., represent the inputs and outputs respectively of the black boxes in equilibrium. All black boxes other than the "controlled system" respond with negligible time delay, but the controlled system has a relaxation time τ_0, the intrinsic relaxation time. If the amplification factor

$$K = \alpha_D \frac{\varepsilon_\infty + 2}{3} \qquad (2.68)$$

is variable, the system changes with increasing K from a stable condition to one of infinite response and infinite relaxation time. The system in question is analogous not only to an electrical network or mechanical system but also to a chemical or nuclear chain reaction.

REFERENCES

1. M. Abraham and R. Becker (1932). "The Classical Theory of Electricity and Magnetism." Blackie & Son, London.
2. J. C. F. Boettcher (1952). "Theory of Dielectric Polarization." Elsevier, Amsterdam.
3. H. Fröhlich (1958). "Theory of Dielectrics." Oxford University Press, London.
4. P. Debye (1929). "Polar Molecules." The Chemical Catalog Co., Inc. Also (1945). Dover Publications, London.
5. H. F. Olson (1958). "Dynamic Analogies", p. 210. D. van Nostrand, London.

CHAPTER 3

The Thermodynamics of Relaxation

A. EQUILIBRIUM CONSIDERATIONS

A relaxation process consists in the approach to equilibrium of a system which is initially out of thermodynamic equilibrium. Such a process is irreversible unless it is infinitely slow, that is it is connected with an increase of entropy in the system and its surroundings which can only be undone at the expense of work. Irreversible processes are not covered by equilibrium thermodynamics, but in certain conditions rates of irreversible change of a system can be deduced from the extent to which an initial state differs from equilibrium. The discipline[1] concerned with this deduction is called "irreversible thermodynamics" or "steady state" thermodynamics, the latter because it is eminently suitable for the treatment of conditions of steady flow. Most relaxation processes fulfil the conditions for which irreversible thermodynamics is applicable. The thermodynamic treatment illuminates the significance of some of the concepts of relaxation, and shows, in particular, that the Debye equations are the most plausible description of a relaxation process in first approximation.

In stable thermodynamic equilibrium any system can be described by a set of independent variables which refer to the system as a whole, and a correspondent set of dependent variables which are uniquely determined by the independent ones[2]. For a perfect gas, independent variables might be, for instance, p and T, and dependent ones the volume v, the internal energy U, the entropy S, the free energy $U - TS$, and other combinations of the above. The definition of all these quantities implies that they can be formulated within narrow limits of uncertainty. The applicability of steady state (or irreversible) thermodynamics implies that all the variables in question can still be meaningfully defined out of equilibrium.

Relaxing systems have normally more than two independent variables. An isothermally expanding gas, with only two independent variables, is a relaxing system, but a rather trivial one. Normally such systems contain some source of energy other than that which appears in the elastic properties as a function of pressure, volume and temperature. This source of energy may be contributed by polarization in an electric or magnetic field. Alternatively it may be some chemical energy, the latter being defined in a very general sense.

For instance the "chemical" energy may represent some change of the structure of a solid which is on a larger than molecular scale and does not respond elastically, as in the case of creep in solids.

In the following we shall treat the electrical case and shall also consider analogies with systems in which the extra source of energy is a chemical reaction. This is interesting because of the light it throws on dielectric relaxation, and also because it links up with the complementary discipline of ultrasonic relaxation.

Before discussing time effects in relaxing systems we shall first discuss two simple cases which show how the electric field enters into the conditions of stable thermodynamic equilibrium. At first, we shall treat a macroscopic system. Following Fröhlich[3] we shall assume a dielectric system of constant volume in contact with a large thermal reservoir of temperature T. In internal equilibrium, this system can be described by its energy content U and its entropy S, and it will be in equilibrium at the temperature T if its Helmholtz free energy

$$F = U - TS \tag{3.1}$$

is a minimum with regard to any possible small change of the independent variables present.

The system is subject to an applied field E, and the energy contributed by this field alters S as well as U, relative to their value in the absence of a field. The extent of this alteration depends on the way in which the system interacts with the large reservoir surrounding it. We shall assume a reversible isothermal change, such that any effect which the field could have on the temperature of the system, if this were isolated, leads to an infinitesimally slow transfer of heat to or from the large reservoir.

We assume a small change of E, while D is assumed to be a unique function of E, namely

$$D = \varepsilon_s(T)E. \tag{3.2}$$

The change of field strength causes an energy

$$\frac{1}{4\pi} E \, dD$$

to enter into the system, while at the same time it alters the entropy by a small amount dS. This change of entropy implies in the conditions assumed that a quantity of heat

$$dQ = T \, dS \tag{3.3}$$

3. THE THERMODYNAMICS OF RELAXATION

is exchanged with the large reservoir. The total change of heat content of the dielectric system is therefore

$$dU = dQ + \frac{E}{4\pi} dD = T dS + \frac{E}{4\pi} dD. \qquad (3.4)$$

Fröhlich calculates S and U as a function of the field. It is convenient to use E^2 as independent variable, and Fröhlich uses a purely macroscopic treatment, based on the fact that dS and dU are total differentials. He finds

$$S = S_0(T) + \frac{\partial \varepsilon_s}{\partial T} \frac{E^2}{8\pi} \qquad (3.5)$$

$$U = U_0(T) + \left(\varepsilon_s + T\frac{\partial \varepsilon_s}{\partial T}\right) \cdot \frac{E^2}{8\pi} \qquad (3.6)$$

while the free energy F (equation 3.1) is given by

$$F = F_0(T) + \frac{\varepsilon_s E^2}{8\pi}. \qquad (3.7)$$

The result shows that the electrostatic energy $\varepsilon_s E^2/8\pi$ stored in the dielectric is a free energy, not an internal energy.

Referring back to the models discussed earlier, they all give (equations 2.44, 2.56)

$$\varepsilon_s = \varepsilon_\infty + \frac{C}{T}. \qquad (3.8)$$

C a constant, so that

$$\frac{\partial \varepsilon_s}{\partial T} = -\frac{1}{T}(\varepsilon_s - \varepsilon_\infty). \qquad (3.9)$$

Substitution of equation (3.9) into equations (3.5) and (3.6) gives

$$U = U_0(T) + \frac{\varepsilon_\infty E^2}{8\pi} \qquad (3.10)$$

$$S = S_0(T) - \frac{\varepsilon_s - \varepsilon_\infty}{T} \cdot \frac{E^2}{8\pi}. \qquad (3.11)$$

Equation (3.11) shows that the entropy decreases with increasing field strength. The term TS which enters into the free energy (equation 3.1) is an energy term. The fact that it is negative signifies that the applied field does work to reduce the entropy, in other words, to create order. Of the electrostatic energy (equation 3.7) only the part proportional to ε_∞ is stored as an internal energy, the rest is used to order the assembly. The polarization of an assembly of dipoles signifies that they are ordered as compared with the unpolarized state.

Equation (3.8) does not apply to dielectrics in general. In fact ε_s may increase with temperature, and in that case most of the free energy of the field may be stored in a U rather than a TS term. This means that the field operates against some other ordering force in the system, generally of chemical origin. The role of the electrical field and of the entropy can be further clarified with the help of a microscopic treatment of the free energy. For this purpose the most convenient model is the bistable model of the two potential wells which we shall consider next.

The bistable model is analogous to a chemical reaction the affinity of which is fully or partially produced by an electric field. The equilibrium thermodynamics of chemical reactions is widely used in chemistry[2], and the relaxation of a chemical system is treated in the theory of rate processes. Relaxations caused by the perturbation of chemical equilibria by pressure changes have been treated in the theory of ultrasonics, and perturbations by an electric field may be treated in an analogous way. We shall here follow the treatments used by Herzfeld and Litovitz[5], and Davies and Lamb[6].

In general, a chemical reaction may be described by

$$\sum \gamma_i M_i = 0 \tag{3.12}$$

where M_i is the molecular weight and γ_i the mole number of the ith constituent. The model in question is analogous to a reaction:

$$A \rightleftharpoons B$$

where one chemical species, A, is converted into another, B, of equal molecular weight but different constitution. Here $M_1 = M_2$ and N_1 units are in form A, with a potential energy as represented by well 1, and N_2 units in form B or well 2 while $N_1 + N_2 = N$, the total number of molecules or double potential wells. In order to conform to equation (3.12) we define fractional number of occupation

$$\gamma_1 = \frac{N_1}{N_1 + N_2} - \frac{1}{2}; \quad \gamma_2 = \frac{N_2}{N_1 + N_2} - \frac{1}{2} \tag{3.13}$$

so that

$$\gamma_1 + \gamma_2 = 0. \tag{3.14}$$

The numerical value

$$|\gamma_1| = |\gamma_2| = \xi \tag{3.15}$$

a dimensionless number, gives the deviation of the system from equal occupation of the two states. It may be considered as a degree of order, or in

3. THE THERMODYNAMICS OF RELAXATION

the chemical sense, as a degree of reaction. For the latter case it would be more usual to count the degree of reaction from the condition when all molecules are in state A, when $\xi = 1$ would represent complete conversion from A to B, but the definition of ξ by equations (3.14) and (3.15) is more convenient here. A general definition of the degree of advancement of a chemical reaction is given by de Groot[4].

In an electric field E acting on a charged particle in a potential well (or dipole of moment μ), a deviation from average occupancy ξ implies a potential energy $2\xi\mu E$ (we are for the moment neglecting electrostatic interaction). This potential energy adds (or subtracts from) any energy difference that may exist between the two wells for reasons other than the electric field. In general, the potential difference between the two wells is

$$U = B \pm 2\xi\mu E \tag{3.16}$$

where B represents an energy of chemical or other origin and will in the following be considered constant.

The entropy of the system containing N_1 dipoles or molecules in state A and N_2 in state B can be calculated statistically. The binomial coefficient $\binom{N}{N_1}$ gives the number of distinct ways in which N entities can be arranged in two groups of N_1 and N_2. The entropy of mixing the two groups is proportional to the logarithm of the binomial coefficient so that

$$S = k \log \frac{N!}{N_1! N_2!} \tag{3.17}$$

where k is the Boltzmann constant. If N_1 and N_2 are large the entropy of mixing can be expressed as a function of the concentrations $\frac{N_1}{N}$ and $\frac{N_2}{N}$. Equation (3.17) may be transformed[2] into

$$S = k\left(N_1 \log \frac{N}{N_1} + N_2 \log \frac{N}{N_2}\right)$$

and introduction of the order variable ξ transforms this expression further into

$$S = -kN((\tfrac{1}{2} + \xi) \log (\tfrac{1}{2} + \xi) + (\tfrac{1}{2} - \xi) \log (\tfrac{1}{2} - \xi)). \tag{3.18}$$

This entropy is always positive since mixing implies disorder. The entropy of mixing is the only relevant entropy term in the case of the bistable model. In the case of a chemical reaction one would also have to consider the difference of entropy of the forms A and B.

The free energy, as defined by equations (3.1), (3.16) and (3.18) permits the calculation of the equilibrium degree of order ξ_e by the condition that the free energy is a minimum for $\xi = \xi_e$

$$\left(\frac{\partial F}{\partial \xi}\right)_T = 0. \tag{3.19}$$

The calculation is very simple for the symmetrical double well ($B = 0$) when $\xi \ll 1$ may be assumed.

For that case

$$\frac{F}{N} = 2\xi\mu E + kT \log 2 - 2kT\xi^2 \tag{3.20}$$

and the equilibrium degree of order is

$$\xi_e = \frac{\mu E}{2kT}. \tag{3.21}$$

This agrees with the result obtained in Chapter 2 (equation 2.38) with the aid of statistical assumptions, since

$$P_s = 2\mu\xi_e N \tag{3.22}$$

$$= \frac{N\mu^2 E}{kT}. \tag{3.22a}$$

Introduction of the equilibrium value ξ_e into equation (3.20) gives for the change of free energy due to the field

$$F(\xi_e) - F(0) = \Delta F = \frac{N\mu^2 E^2}{2kT}. \tag{3.23}$$

Translated into macroscipic terms, using equation (3.22a) this gives

$$\Delta F = (\varepsilon_s - \varepsilon_\infty)\frac{E^2}{8\pi} \tag{3.24}$$

in view of equations (2.13) and (2.14). Comparison of equations (3.24) and (3.11) confirms that the contribution of the bistable dipoles is entirely an entropy term.

The thermodynamic treatment of the bistable system can be extended to the case where B in equation (3.16) is not negligible. The result has already been given in equation (2.48). The treatment can also be extended to the case of a mutual interaction between dipoles. An exact treatment for this case is extremely complex, since it depends on the geometry and size of the array[3]. However, the simplified treatment discussed in the second chapter illustrates the physical meaning.

For the bistable model where electrostatic interaction acts so that the field acting on a dipole is increased to E_i according to equation (2.57), and where $\varepsilon_\infty = 1$, we may calculate the internal energy as

$$U = \int_0^P E_i \, dP \tag{3.25}$$

$$= \int_0^P \left(E + \frac{4\pi}{3} P\right) dP$$

$$= 2\mu N \int_0^\xi \left(E + \frac{8\pi}{3} \mu \xi N\right) d\xi$$

$$= 2\xi \mu N E + 2\xi^2 \frac{4\pi}{3} \mu^2 N^2 \tag{3.25a}$$

while the entropy is given by equation (3.18) for $\xi \ll 1$.

The free energy is now given by

$$\frac{F}{N} = 2\xi \mu E + kT \log 2 - 2\xi^2 \left(kT - \frac{4\pi}{3} \mu^2 N\right). \tag{3.26}$$

The equilibrium value ξ_e for this model may be unequal zero for $E = 0$, in which case there exists a spontaneous polarization (see Chapter 16, Section D). When there is no spontaneous polarization, the equilibrium degree of order follows by differentiation of equation (3.26), as

$$\xi_e = \frac{\mu E}{2kT} \cdot \frac{1}{1 - \frac{4\pi \mu^2 N}{3kT}}$$

$$= \frac{\mu E}{2kT} \cdot \frac{T}{T - T_c} \tag{3.27}$$

where

$$T_c = \frac{4\pi}{3} \cdot \frac{\mu^2 N}{k}. \tag{3.28}$$

The condition that there is no spontaneous polarization implies that $T > T_c$.

Equation (3.26) is interesting in that it shows how electrostatic interaction works: it reduces the entropy term which opposes the ordering of the array of dipoles.

B. RELAXATION IN TERMS OF IRREVERSIBLE THERMODYNAMICS

Equilibrium thermodynamics does not by itself give information or relaxation. The presence of a free energy ΔF due to the field in a dielectric means that the system is able to do that much useful work. If it does, then we do not speak of relaxation. For instance the model just discussed could be used as a means for adiabatic cooling, if on removal of the field it would cool down so that the cooling corresponded to the limiting thermodynamic efficiency.

Relaxation occurs when the free energy stored in the system is degraded to heat; in other words, if entropy is created irreversibly. In the system envisaged this means that on removing the field some or all the free energy is dissipated in the dielectric without reappearing in a corresponding adiabatic cooling. The free energy of the field has been used to increase the total amount of heat stored in the dielectric plus the heat reservoir surrounding it.

Irreversible thermodynamics is concerned with the rate at which entropy is created in a system. This rate depends on the extent to which the system is out of equilibrium, and can be calculated as a function of the deviation, provided that certain conditions are fulfilled. A more detailed discussion of these conditions will follow in Chapter 4, but one of their essentials is that the deviation from equilibrium should be small, and thermodynamic functions in the system still defined within close limits. We shall apply the methods of irreversible thermodynamics to dielectrics and see that they lead to the right results.

Irreversible thermodynamics operates with a system of forces and fluxes[1]. A "force" X characterizes the deviation of the system from equilibrium, while the "flux" Y represents a flow or other change in the system which causes it to move towards equilibrium. A system may contain several linked forces and fluxes X_i and Y_i, and the sum of the scalar products of fluxes and forces gives the rate of creation of entropy by the process in question as

$$\sum_i T \frac{dS}{dt} = \sum_i X_i Y_i. \qquad (3.29)$$

A product of T and $\frac{dS}{dt}$ is an energy dissipation.

For the bistable system $i = 1$ the relevant flux is the change of the dimensionless "degree of order" or "degree of reaction" parameter ξ, that is

$$Y = \frac{d\xi}{dt}. \qquad (3.30)$$

The "force" in question is the deviation of the free energy from its equilibrium value. For the free energy F used so far, we define

$$X = A = \left(\frac{\partial F}{\partial \xi}\right)_{T, E} \qquad (3.31)$$

the letter A being used for chemical affinity. For the simplest version of the bistable system this gives (equation 3.20)

$$A = 2(\mu E - 2\xi kT) \tag{3.32}$$

which is zero if ξ is in equilibrium with the field E, but has a finite value otherwise.

If steady state thermodynamics applies we expect a linear relationship between force and flux, so that

$$\frac{d\xi}{dt} = \frac{1}{K} \cdot A \tag{3.33}$$

where K is constant with respect to time. Equations (3.32) and (3.33) lead obviously to the linear differential equation of a relaxation process (1.31) whose relaxation time is given by

$$\tau = \frac{K}{4kT}. \tag{3.34}$$

Since ξ is proportional to P_s according to (3.22) an equation with the same τ holds for P also. The value of K cannot be determined without recourse to molecular quantities. A modified τ applies if we consider the electrostatic interaction, as may be deduced from (3.26).

The argument so far neglects the effect of compression on the equilibrium. This is permissible in most (though not all) dielectric phenomena, but the connection between ultrasonic and electrical relaxation becomes apparent only if we take the $p.v.$ term into account. A more general formulation of the relaxation of a chemical equilibrium uses the thermodynamic potentials[2,5]

$$dH = T\,dS + v\,dp + A\,d\xi \tag{3.35}$$

or[6]

$$dG = S\,dT + v\,dp + A\,d\xi \tag{3.36}$$

where ξ is some generalized order parameter. In either case a treatment analogous to that given above, together with an expansion of A as a function of ξ, leads to

$$\frac{d\xi}{dt} = K\frac{(\partial A)}{(\partial \xi)}(\xi - \xi_0) \tag{3.37}$$

so that

$$\frac{1}{\tau} = K\frac{\partial A}{\partial \xi}. \tag{3.38}$$

Equations (3.37) and (3.38), which are a more sophisticated formulation of equations (3.33) and (3.34), imply that the relaxation time may be different according to which variables are held constant during a given experiment. In general

$$\left(\frac{\partial A}{\partial \xi}\right)_{T,p} \neq \left(\frac{\partial A}{\partial \xi}\right)_{T,v} \qquad (3.39)$$

so that the relaxation time is different for conditions of constant pressure and constant volume. The inter-relations between the various partial differential quotients which may be relevant are reviewed by Herzfeld and Litovitz[5].

In the simple electrical case treated above A was due entirely to the field. In general, A may be due partly to other causes, and only be influenced by the field. For ultrasonic relaxation the origin of A is chemical, in a general sense, and ξ is perturbed by a change of pressure.

The thermodynamic treatment shows that the dielectric and ultrasonic relaxation times should be identical, or very similar, if it is the same ξ that is perturbed in one case by an electric field and in the other by a pressure. Differences might be caused by different variables being kept constant in the process in question, or by a variation in the role of electrostatic interaction. However, these differences would not normally affect the order of magnitude of τ. In the experimental section we shall see that similar relaxation times follow quite often from dielectric and ultrasonic measurements. However, there are cases when an electric field cannot perturb an equilibrium because no electric dipoles are present, while a pressure has effect. *Vice versa* volume effects connected with dielectric relaxation may be so small that pressure changes have no effect on the relevant order variable. In general, it does not follow that dielectric and ultrasonic relaxation processes need be connected.

C. THE DISSIPATION OF ENERGY

For a relaxation process where a degree of order can be defined the rate of dissipation of energy (equations 3.29–3.31) is given by

$$T\frac{dS}{dt} = A\frac{d\xi}{dt}. \qquad (3.40)$$

For the bistable system treated we can follow the entropy changes in detail. Substituting from equations (3.32) and (3.33) for the simplest model, we have

$$T\frac{dS}{dt} = \frac{4}{K}(\mu E - 2\xi kT)^2 \qquad (3.40a)$$

giving the irreversible heating as a function of $\xi(t)$. For a step-function where

3. THE THERMODYNAMICS OF RELAXATION

$E = E_0$ for $t < 0$, and zero for $t > 0$, equation (3.33) may be integrated to give (see equation 3.32)

$$\xi = \frac{\mu E_0}{2kT} e^{-\frac{t}{\tau}}. \qquad (3.41)$$

Introduction of ξ into equation (3.40) shows that the entropy increases monotonously. If equation (3.40) is integrated it is found that

$$T \int_0^\infty \frac{dS}{dt} dt = \frac{N\mu^2 E_0^2}{2kT}. \qquad (3.42)$$

This is exactly the total free energy stored in the system on application of the field (equation 3.23). That is, in the process considered (which implies constant temperature) all the free energy of the field is irreversibly dissipated into heat energy. Alternatively, it may be said that all the negative entropy contributed by the field is dissipated until it is completely wasted.

The above treatment of the bistable model was carried out in terms of the order variable ξ. This was done in order to illustrate the significance of some concepts, but in general a much simpler method may be used. The bistable model is a special case of a Debye dielectric with $\varepsilon_s - \varepsilon_\infty = \frac{N\mu^2}{2kT}$ and $\varepsilon_\infty = 1$. A Debye dielectric has in general an equivalent circuit as signified in Chapter 1, Fig. 4, where $C_1 V = \varepsilon_\infty E$ and $C_2 V = (\varepsilon_s - \varepsilon_\infty)E$, and this circuit responds electrically exactly like the dielectric.

Thermodynamics is no respecter of models and the generation of entropy for a given process may be derived for the equivalent circuit, using equation (3.29) so that[1] the force X is the voltage and the flux Y is the current

$$T \frac{dS}{dt} = V(t) . I(t). \qquad (3.43)$$

We shall consider the behaviour of a Debye dielectric subjected to alternating current in terms of its equivalent circuit.

In calculating the product in equation (3.43) it is convenient to use complex quantities. In terms of such quantities the real current and voltage are in general[7] given by

$$V(t) = \tfrac{1}{2}(V e^{i\omega t} + \bar{V} e^{-i\omega t}) \qquad (3.44)$$

$$I(t) = \tfrac{1}{2}(I e^{i\omega t} + \bar{I} e^{-i\omega t}) \qquad (3.45)$$

where V and $\bar{V}$ are complex conjugates and I and $\bar{I}$ likewise. For the purposes of calculating the product we may put $V = \bar{V} = |V| = V_0$ while

$$I = I_0 e^{i\phi}, \qquad \bar{I} = I_0 e^{-i\phi} \qquad (3.46)$$

where $\phi = \dfrac{\pi}{2} - \delta$ is the phase angle between current and voltage.

Introduction of equations (3.44)–(3.46) into equation (3.43) gives

$$T\frac{dS}{dt} = V_0 I_0 (e^{i\phi} + e^{-i\phi}) + V_0 I_0 (e^{i(2\omega t - \phi)} + e^{-i(2\omega t - \phi)})$$

$$= \tfrac{1}{2} V_0 I_0 \cos \phi + \tfrac{1}{2} V_0 I_0 \cos (2\omega t - \phi). \qquad (3.47)$$

The first term is time independent while the second term depends periodically on time. The significance of the second term may most easily be seen if we consider a pure capacitance where $\phi = \pi/2$ and the first term vanishes. Since we have in general

$$V_0 I_0 = V_0^2 \cdot \frac{1}{Z(\omega)} \qquad (3.48)$$

where $Z(\omega)$ is the modulus of the complex impedance and since $Z(\omega) = 1/\omega C$ for a pure capacitance equation (3.47) transforms in the given case into

$$T\frac{dS}{dt} = -\tfrac{1}{2}\omega C V_0^2 \sin 2\omega t. \qquad (3.49)$$

This equation shows that the generation of entropy is periodic with twice the period of the applied field. Negative entropy is produced while the capacitor is being charged and positive entropy when it is being discharged. In the average over the period of a cycle

$$\int_0^\omega T\frac{dS}{dt} = 0 \qquad (3.50)$$

and this applies in general to the second term in equation (3.47). The second term does not represent a net generation of entropy.

The first term in equation (3.47) is always positive, and it represents a generation of entropy at a constant rate. It may be understood[7] as the ohmic heating in the resistance R_2 of the equivalent circuit. This term may be written in the form (1.27) for the equivalent circuit. For the Debye dielectric the energy dissipation in practical units may be written in the form

$$T\frac{dS}{dt} = \tfrac{1}{2} E_0^2 \varepsilon_0 (\varepsilon_s - \varepsilon_\infty) \frac{\omega^2 \tau}{\sqrt{(1 + \omega^2 \tau^2)}} \qquad (3.51)$$

where E_0 is the peak value of the electric field. For $\omega = 0$ the irreversible generation of entropy is zero. This implies that the polarization is reversible for an infinitely slow application of the field, as has been anticipated. For infinite ω the power loss is finite, and a maximum, namely (see equation 2.26)

$$T\frac{dS}{dt} = \frac{E_0^2}{2} \cdot \frac{\varepsilon_0(\varepsilon_s - \varepsilon_\infty)}{\tau}. \tag{3.52}$$

In terms of the circuit analogue (1.4) this signifies that for finite frequencies the series capacitance C_2 reduces the magnitude of the current flowing through branch 2, while for infinite frequencies C_2 acts as a short circuit. In terms of dielectric theory the increase of the power loss with frequency means that for zero frequency the response is entirely due to spontaneous fluctuations, while for higher frequencies the spontaneous character of the fluctuations is increasingly perturbed. This will be discussed in the next chapter.

REFERENCES

1. K. G. Denbigh (1950). "The Thermodynamics of the Steady State." Methuen, London.
2. R. Fowler and Guggenheim (1939). "Statistical Thermodynamics." Cambridge University Press, London.
3. H. Fröhlich (1958). "Theory of Dielectrics." Oxford University Press, London.
4. S. R. de Groot (1958). "Thermodynamics of Irreversible Processes." North Holland Publications, Amsterdam.
5. K. H. Herzfeld and T. A. Litovitz (1959). "Absorption and Dispersion of Ultrasonic Waves." Academic Press, New York and London.
6. R. O. Davies and J. Lamb (1957). Q. Rev. chem. Soc., **2**, 134.
7. E. A. Guillemin (1953). "Introductory Circuit Theory." John Wiley and Chapman & Hall, London.

CHAPTER 4

Fluctuations and Dielectric Relaxation

A. EQUILIBRIUM AND FLUCTUATIONS IN A MICROSCOPIC ASSEMBLY

In Chapters 2 and 3 the exposition proceeded from the definition of a state of equilibrium to a definition of relaxation as a progress towards equilibrium. When one considers a microscopic assembly this sequence is no longer obvious. For such an assembly it is easier to grasp the existence of forces which cause relaxation than it is to define a state of equilibrium in an exact manner.

A microscopic assembly which consists of atoms, molecules or similar small entities differs in kind from an assembly which consists of macroscropic objects, say of billiard balls. The microscopic particles are spontaneously in motion in an irregular way which is seen in daily life only rarely, as for instance in a swarm of insects.

The behaviour of microscopic assemblies is most easily visualized for a gas where atoms (or molecules) move freely through space. In a gas each atom has at a given time a given velocity in a given direction and a corresponding kinetic energy. The gas is capable of a state of equilibrium because atoms collide and interchange energy and momentum. It is easy to see that if all atoms in the left half of a container had larger kinetic energies than all atoms in the right half, collisions between atoms would cause a transfer of kinetic energy from left to right. The assembly assumed is clearly not in equilibrium and will in time change towards a state closer to equilibrium. However, it is not easy to give exact definitions for a state of equilibrium for an assembly where atoms move about spontaneously and collide from time to time.

Equilibrium in microscopic assemblies is treated in text books. The basic, and by no means easy step, is the definition of temperature[1]. In equilibrium, each particle has in the average an energy $\frac{1}{2}kT$ for each degree of freedom. Here T is the absolute temperature counted from absolute zero and k is the Boltzmann constant. The number of degrees of freedom is the number of independent variables necessary to describe uniquely the spatial arrangement of a system. A simple atom in a gas has three degrees of freedom corresponding to the components of its velocity in the x, y and z direction. A molecule has in addition internal degrees of freedom corresponding to vibrations and rotations. Apart from the temperature statistical thermodynamics[1,2] defines a number of other macroscopic variables which are also attributes of the assembly as a whole, in equilibrium. Among these are energy, density, electrical or magnetic polarization, etc.

From the viewpoint of relaxation it is essential to realize that all thermodynamic variables are averages. This implies the existence of fluctuations. For instance the kinetic energy of individual gas atoms differs very greatly even though each atom has in the average an energy $\frac{3}{2}kT$. It is an essential feature of statistical thermodynamics that it defines probabilities for the possession of a certain attribute such as energy by a given particle. This may be expressed in terms of a distribution function. For instance, for a gas containing N atoms the number $n(c)$ of atoms which have velocities (in any direction) between c and $c + dc$ is given by

$$n(c)\,dc = 4\pi N c^2\,dc\left(\frac{m}{2\pi kT}\right)^{\frac{3}{2}} e^{-\frac{1}{2}\frac{mc^2}{kT}} \tag{4.1}$$

where m is the mass of the atom.

Equation (4.1) means that there is a finite probability that an atom may have a very high kinetic energy even if the average kinetic energy is small. However, this probability is governed by the exponential term which is very small in the case assumed. A probability so defined implies fluctuations in the sense that the average value of c taken over finite times and finite volumes deviates from the average value of the velocity for the assembly as a whole. In the same way, fluctuations occur with regard to pressure and temperature and the other macroscopic variables which describe the assembly in equilibrium.

The most abstract macroscopic variable defined by thermodynamics is the entropy which for a fixed temperature may be defined statistically as the logarithm of the number of alternative ways in which a given state may be realized, multiplied by a factor of dimension kT. The entropy of an infinitely large closed system can only increase in time. When a gas container of a certain volume is brought into communication with an empty container the gas expands spontaneously into the empty portion and the entropy increases. The gas will not return spontaneously into the smaller volume to stay there, it can only be pumped back at the expense of energy supplied to the system from outside. However, microscopic fluctuations of density do occur in a gas in equilibrium and the entropy itself fluctuates in space and time. For our container of finite size there exists a finite, if very small, probability that the gas out of the larger volume will for a finite period congregate in the initial smaller volume. The magnitude of this probability depends not only on the size of the assembly but also on its relationship to the outside world, for instance on whether the gas exchanges heat with an outside reservoir or exists in isolation.

Fluctuations in equilibrium are part of the entropy of the system in which

4. FLUCTUATIONS AND DIELECTRIC RELAXATION

they occur but they do not create entropy. The system carries out its equilibrium fluctuations over periods of any length, at constant average entropy.

When a system is disturbed out of its equilibrium by, say, a sudden change of pressure, when the new pressure demands a modified distribution function, fluctuations tending towards the new equilibrium become more probable than those tending away from it. The resultant asymmetry of the fluctuations provides a "force" which tends to establish equilibrium and causes a "flux". Irreversible thermodynamics applies to cases where the "force" X tending towards equilibrium is supplied by the fluctuations, and where the fluctuations are still close to their equilibrium character.

The proviso that the fluctuations should be not too far removed from their equilibrium character is formulated in irreversible thermodynamics with the help of Onsagers principle of microscopic reversibility, which has been expressed briefly by Tolman[2] as follows:

"Under equilibrium conditions any molecular process and the reverse of the process will be taking place in the average at the same rate."

The molecular processes of dielectrics relaxation concern the movements of dipole or other polarizable elements whose resultant polarization fluctuates about an average. The condition of microscopic reversibility would seem to mean here that a movement in the direction of the field and a movement in the opposite direction should be equally probable, and Kittel[3] points out that this appears to contradict the kinetic derivation of the relaxation process given in our Chapter 2 for the bistable model (equations 2.27–2.48). However, the contradiction is only apparent.

Kittel[3] (Section 39) treats the behaviour of the bistable model for the case that the two energy states are coupled by a weak interaction with lattice oscillators in an associated heat reservoir. For the sake of simplicity, the energy of an oscillator is taken as $hv = \mu E$. In that case the jump of a dipole from 2–1 implies that an oscillator receives the energy μE, while a jump from 1–2 means that an oscillator gives up an energy μE to a dipole (disordering cools the reservoir). We shall not reproduce the calculation here, only point out that the principle of microscopic reversibility is fulfilled if we take as our system not the dipoles alone but the dipoles together with the oscillators.

Kittel points out that similar results follow for any assumptions we may use to describe the nature of the reservoir and the coupling between reservoir and dipoles. All these elaborations of the bistable model lead to the relaxation equation for the polarization provided that $\mu E \ll kT$.

B. FLUCTUATIONS IN THE BISTABLE MODEL IN EQUILIBRIUM

Although the forces tending towards equilibrium are easier to understand than the definition of equilibrium it is still convenient to treat fluctuations on the basis of equilibrium theory. We shall introduce a simple mathematical relationship between the probability of the fluctuations of a macroscopic variable x and the free energy of the assembly in question.

The magnitude of the fluctuation of a quantity x is defined as

$$\overline{(x - \bar{x})^2} = \langle (\Delta x)^2 \rangle = \text{var } x \tag{4.2}$$

where $\bar{x}$ is the average value of x, so that

$$\overline{x - \bar{x}} = 0. \tag{4.3}$$

If the quantity x is one of the macroscopic variables which determine the free energy of a system in statistical mechanics, then the equilibrium value of $\overline{x^2}$ can be determined from the free energy. The free energy is by definition a minimum for the equilibrium value $\bar{x}$ of x, and we define

$$q = x - \bar{x} \tag{4.4}$$

as the deviation of x from equilibrium observed during a short time interval within a finite volume.

The free energy will in general depend not only on x but also on other variables. Relevant variables for a macroscopic system may be, for instance, the pressure, energy content and electric polarization. In order to evaluate the effect of q on the free energy we keep all variables other than x constant and find

$$F(q) = F_0 + q \left(\frac{\partial F(x)}{\partial x} \right)_{q=0} + \frac{q^2}{2} \left(\frac{\partial^2 F(x)}{\partial x^2} \right)_{q=0} + \cdots . \tag{4.5}$$

Since $q = 0$ implies equilibrium, the second term of this expression vanishes by definition and

$$\Delta F = F(q) - F_0 = \frac{q^2}{2} \left(\frac{\partial^2 F(x)}{\partial x^2} \right)_{q=0} \tag{4.6}$$

for small values of q, provided that the second differential does not vanish.

Equation (4.6) connects a fluctuation of x with a fluctuation ΔF of the free energy at a given instant. The average fluctuations of the variable x is an equilibrium property of the system and can be calculated if we make an assumption about the connection between ΔF and the probability function which refers to x. Such an assumption was introduced by Einstein. It consists of treating the free energy of a fluctuation as if it referred to a degree of freedom of the system. This approach has been discussed in some detail by Lax[4]. Here it will be used in an elementary way, by putting

$$p(q) = \text{const. } e^{-\frac{\Delta F(q)}{kT}} \tag{4.7}$$

4. FLUCTUATIONS AND DIELECTRIC RELAXATION

for the probability of finding the system with a deviation q of the variable x from equilibrium, at a given instant. Since $\Delta F(q)$ is known from equation (4.6), equation (4.7) gives directly the probability of a fluctuation of the magnitude defined by q.

When the variable q is replaced by x, combination of equations (4.6) and (4.7) gives

$$p(x) = \text{const.} \; e^{-\frac{(x-\bar{x})^2}{2kT}\left(\frac{\partial^2 F}{\partial x^2}\right)_{x=\bar{x}}} \tag{4.8}$$

This expression is a special case of a normal distribution or Gaussian probability density function which has the general form

$$p(x) = \frac{1}{\sigma\sqrt{2\pi}} e^{-\frac{(x-\bar{x})}{2\sigma^2}} \tag{4.9}$$

The value of the constant before the exponential term follows from the condition that the sum of all probabilities is unity. The constant σ is the standard deviation of the normal distribution.

Comparison of equations (4.8) and (4.9) identifies the standard deviation for the probability of a deviation of x by

$$\sigma^2 = \frac{kT}{\left(\frac{\partial^2 F}{\partial x^2}\right)_{x=\bar{x}}}. \tag{4.10}$$

On the other hand, σ^2 is in general equal to the fluctuation of x since

$$\overline{(x-\bar{x})^2} = \int_{-\infty}^{+\infty} (x-\bar{x})^2 p(x-\bar{x}) \, dx$$

$$= \sigma^2. \tag{4.11}$$

Equations (4.10) and (4.11) show that the fluctuations of a variable can be calculated for any system for which the free energy can be expressed in terms of the fluctuating variable.

For the bistable models without electrostatic interaction, $\varepsilon_\infty = 1$ and $\xi \ll 1$ the free energy/unit volume is given by equation (3.20). For this simplest model the volume or shape of specimen does not affect the equilibrium value of ξ (see Chapter 16, Sections D and E). For the consideration of fluctuations the volume is relevant. The second differential of the free energy of a specimen of volume v may be deduced from equation (3.20) as

$$\left(\frac{\partial^2 F}{\partial \xi^2}\right)_{\xi=\bar{\xi}} = 4NvkT. \tag{4.12}$$

Comparison of equations (4.10), (4.11) and (4.12) shows that

$$\overline{(\xi - \bar{\xi})^2} = \frac{1}{4Nv} \tag{4.13}$$

which means that for a small $\bar{\xi}$ and a small number of fluctuating dipoles the average deviation of ξ from its equilibrium value may be larger than that equilibrium value itself.

The order variable ξ is related to the polarization P of the assembly of bistable dipoles by equation (3.22) while the electric moment of a specimen of volume v is

$$M = Pv \tag{4.14}$$

so that

$$M = 2\mu N v \xi. \tag{4.15}$$

The fluctuation of this moment is given by

$$\overline{(M - \bar{M})^2} = \overline{(\xi - \bar{\xi})^2} \cdot (2\mu N v)^2$$
$$= \mu^2 N v. \tag{4.16}$$

This expression is interesting in that it shows the fluctuation of the electric moment to be independent of the temperature and the field. A small applied field does not influence the magnitude of the fluctuation of M, it only shifts the average value $\bar{M}$.

For the bistable model, as defined, the fluctuation of M can be expressed in terms of the static dielectric constant.

Combination of the general equation (2.10) with equation (2.38) which applies to the model gives

$$\frac{\varepsilon_s - 1}{4\pi} = \frac{\mu^2 N}{kT}. \tag{4.17}$$

Hence the fluctuation of M is

$$\overline{(M - \bar{M})^2} = \frac{1}{4\pi} kTv(\varepsilon_s - 1) \tag{4.18}$$

for the model in question.

The fluctuation of the electric moment of a dielectric can in general be expressed in terms of the dielectric constant, but the relationship is complicated by electrostatic interaction.

This interaction raises great difficulties[6,7] when one wishes to deduce the fluctuation of the polarization of a dielectric specimen from fluctuations on an atomic scale, as will be discussed in Chapter 8. The complications due to

4. FLUCTUATIONS AND DIELECTRIC RELAXATION

electrostatic interaction are less serious insofar as they concern the shape of macroscopic specimens, and the extent to which the external field of a specimen reacts on its polarization. This aspect will be discussed in Section E of Chapter 16.

The fluctuations of the polarization are measurable[8,9]. They manifest themselves as spontaneous fluctuations of the potential between the electrodes of a capacitor containing a dielectric, but sensitive experimental techniques are necessary for their detection. Some measurements will be discussed in Section E of Chapter 16 for ferroelectrics where fluctuations are particularly large[8].

C. THE RATE OF FLUCTUATIONS

So far, we have only considered the average magnitude of fluctuations, but not their dependence on time. This may be described with the aid of suitably defined functions, namely the frequency spectrum $G(f)$ and the autocorrelation function $C(f)$. These two functions represent alternative ways of describing the same phenomenon. They are treated in detail in textbooks dealing with noise and probability theory[9]. Here we shall only give a brief sketch of the mathematical formalism, with a view to application to the simplest dielectric model. We shall follow the concise presentation of Kittel[3] (see Section 28). A more recent and advanced review has been given by R. Kubo[3a].

Figure 1 shows a record of a fluctuating variable $x(t)$ in a system for three different time intervals of moderate length, the average of x being zero. In general a system may change its properties in the time between any two intervals in which fluctuations are observed. However, we shall confine ourselves to processes where it does not, and in particular to a stationary random process for which a single probability function can be constructed so that it fits the records for all time intervals. The relationship between the probability function and the fluctuations may be illustrated by stating that an automatic machine selecting random numbers could be made to produce different records of the type shown, while operating always by the same mechanism. The fundamental concept of randomness which is extremely difficult has been discussed authoritatively by Tolman[2]. In this monograph it will only be used and illustrated in an elementary way.

Given many sets of records as shown in Fig. 1, one may develop each of them into a Fourier series so that

$$x(t) = \sum_{n=1}^{\infty} (a_n \cos 2\pi f_n t + b_n \sin 2\pi f_n t) \qquad (4.19)$$

where f_n is the nth part of the duration of the record (the records are assumed of equal length). As between the short records shown the coefficients a_n and

b_n may vary widely. In the average over all the records they will be distributed in a way which is characteristic for the random process in question.

While the average $\overline{x(t)}$ is assumed zero in the time average over all records the average square term $\overline{x(t)^2}$ will be unequal zero. In the literature the term $\overline{x(t)^2}$ tends to be called a power loss since it is often closely related to an

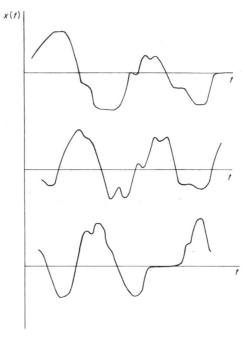

FIG. 1. Three short records of the random fluctuation in time of a variable x, thought to be selected at random out of a long recording. The average value of x over an indefinitely long period is zero.

energy term which may be dissipated. However, the term need not involve a dissipation. The random process in question may proceed at constant entropy.

In order to analyse the distribution of the terms a_n and b_n we calculate a power term W_n for each of them. This is convenient because in $\overline{x^2(t)}$ all cross terms $a_n a_m$, $a_n b_m$, etc. vanish in the average. Hence we put

$$x^2(t) = \sum_n W_n \qquad (4.20)$$

where

$$W_n = (a_n \cos 2\pi f_n t + b_n \sin 2\pi f_n t)^2. \qquad (4.21)$$

4. FLUCTUATIONS AND DIELECTRIC RELAXATION

It can be shown[3] that for a Gaussian random process W_n can be expressed very simply. In the average over time and a very large number of records

$$W_n = \overline{a_n^2 \cos^2 2\pi f_n t} + \overline{b_n^2 \sin^2 2\pi f_n t}$$

$$= \frac{\overline{a_n^2}}{2} + \frac{\overline{b_n^2}}{2} = \overline{a_n^2} = \overline{b_n^2} = \sigma_n^2 \quad (4.22)$$

where σ_n^2 is the standard deviation of the normal distribution of x for the particular frequency which is characterized by the index n.

The magnitude of σ_n^2 as a function of frequency defines the power spectrum for the random process or "noise" in question. We define a function $G(f)$ as the spectral density in a narrow frequency band Δf_n by

$$G(f_n) \Delta f_n = \sigma_n^2 \quad (4.23)$$

The quantity σ_n^2 is measurable. In circuitry the noise which is a feature of the current through, or voltage across, a noisy component may be measured by using a receiver incorporating a filter, passing only frequencies within a narrow band Δf_n. The term σ_n^2 then signifies the power of the noise emitted within the frequency range stated.

The sum of the contributions from all frequencies gives the total average power of the random process, according to

$$\overline{x^2(t)} = \sum_n G(f_n) \Delta f_n \quad (4.24)$$

and the sum may be replaced by an integral for a continuous distribution. The average is independent of time.

The power spectrum $G(f)$ is not the only way of describing the frequency dependence of noise. An alternative method is the definition of an autocorrelation function. The two functions defined are related to one another by the Wiener–Khintchine theorem.

The auto-correlation function is defined as

$$C(t) = \overline{x(a)x(a + t)} \quad (4.25)$$

where the average is taken over all times a and all records. $C(t)$ describes how far the fluctuations change, in the average, after a lapse of time t. This function may be calculated using the Fourier expansion (4.19) and is found to be

$$C(t) = \sum_{n=1}^{\infty} \sigma_n^2 \cos 2\pi f_n t \quad (4.26)$$

or, in view of (4.23),

$$C(t) = \sum_{n=1}^{\infty} G(f_n) \cos (2\pi f_n t) \Delta f_n.$$

For a continuous distribution this becomes

$$C(t) = \int_0^\infty G(f) \cos(2\pi f t) \, df. \tag{4.27}$$

Equation (4.27) shows $G(f)$ and $C(t)$ to be related, in that $C(t)$ is the Fourier cosine transform of $G(f)$. The inverse transform gives $G(f)$ as a function of $C(t)$ by

$$G(f) = 4\int_0^\infty C(t) \cos(2\pi f t) \, dt. \tag{4.28}$$

The two functions $G(f)$ and $C(t)$ are equally suitable to describe the behaviour of fluctuations but one or the other may be more easily accessible for a particular problem.

The auto-correlation function is very simple for a Markoff process, which is a type of stationary random process governed by a Gaussian probability density function. Such a process[9] is one where the state of the system at a given time influences only the state immediately following the present state, and not the more distant future. For such a process it can be shown[9] that

$$C(t) = Q\, e^{-\frac{|t|}{\tau}} \tag{4.29}$$

where Q and τ are constants and $|t|$ indicates that time may be reckoned either as future or past. The constant τ which is called a regression time may be identified with the relaxation time of a relevant process.

D. THE STATISTICAL SIGNIFICANCE OF THE RELAXATION TIME FOR THE BISTABLE MODEL

In the following, the mathematical formalism sketched in Section 4C will be applied to the bistable model. To this end the auto-correlation time will be derived, following the treatment given by Bendat[9] for binomial and Poisson distributions and by Rice[10] for the auto-correlation function of random telegraph signals. It will be shown that a function of form (4.29) results, where τ now has a significance in terms of the bistable model.

In Chapter 2 the dynamic behaviour of the bistable system was defined in the sense that a charge may be located in one or the other of two potential wells and that occasionally it jumps from one well to the other, over a potential barrier. In the absence of a field the number of jumps/unit time is w, in either direction.

In order to derive the auto-correlation function we have to consider the model in statistical terms. We now state that the charge makes a large number of attempts to jump for every successful jump. The physical significance of the attempts made will be discussed in the next section. Here it is essential that

4. FLUCTUATIONS AND DIELECTRIC RELAXATION

successes and failures follow each other at random, so that the result of one attempt does not affect the next one. For the limit of an infinitely large number of tries we may define a probability p as the ratio of successes to the number of tries; for a finite number of tries the number of successes fluctuates.

We consider a number N of tries and ask how likely it is that n out of N tries will succeed. If the sequence of successes and failures were fixed the probability would be $p^n(1-p)^{N-n}$, since the probability of a failure is $1-p$. However, the sequence of successes and failures within the N tries is immaterial. Hence the above probability is multiplied by a binomial coefficient which gives the number of ways in which N items can be arranged in two groups numbering N and $N-n$ respectively. In consequence the probability of n successes out of N attempts is

$$f(n) = \frac{N!}{n!(N-n)!} p^n .(1-p)^{N-n}. \tag{4.30}$$

The function $f(n)$ is a probability density function and obeys the condition

$$\sum_{n=1}^{N} f(n) = 1. \tag{4.31}$$

It has a maximum for a value of n given by

$$\bar{n} = \sum_{n=1}^{N} nf(n) \tag{4.32}$$

and it can be shown[9] that

$$\bar{n} = Np \tag{4.33}$$

as predicted by common sense. However, for series of finite numbers of tries the number of successes fluctuates. If we arrange for m series of N tries each we shall find various numbers $n_1, n_2, \ldots n_m$ for the numbers of successes within a given series.

In order to derive the auto-correlation function by the method used by Rice[10] we consider a given double well and observe it for a long time, as the number of successful jumps fluctuates in the manner described. We now assume that $p \ll 1$ and that N is a large number. Furthermore, we introduce time into the consideration by putting

$$wt_0 = Np \tag{4.34}$$

for the average, most probable, number of successes within a series of N tries, where t_0 is the duration of a series of N tries and w the number of successes (jumps)/unit time. This implies that the tries are successive. We assume that tries occur at random, so that the events within a time interval Δt do not influence the events within any other time interval. With these assumptions equation (4.30) may be transformed[9] into

$$f(n) = \frac{(wt_0)^n}{n!} e^{-wt_0}. \qquad (4.35)$$

This expression is a Poisson probability density function. It may be transformed further into a Gaussian probability density function (see equation 4.11) if n is treated as a continuous variable and w is large. In that case the mean value and standard deviation of the resulting Gaussian distribution are given by[9] $\bar{n} = wt_0$ and $\sigma = \sqrt{(wt_0)}$.

The auto-correlation function may now be derived elegantly as follows[10]. The individual dipole under observation has at any time a moment of either $x(t) = +\mu$ or $x(t) = -\mu$ according to whether it is in well 1 or 2. The auto-correlation function $C(t)$ is the average value of the product $x(a) \times (a + t)$ according to equation (4.25). For any given two values a and t the product is either $+\mu^2$ if the dipole points in the same direction at the two instants chosen, or $-\mu^2$ if the dipole has changed sign an uneven number of times between a and $a + t$. The average of the product is derived by way of a summation over all possible values which a may take up. The average value of $x(a) \times (a + t)$ is given by

$+\mu^2$ (probability of an even number of changes of sign in the interval between a and $a + t$).
$-\mu^2$ (probability of an odd number of changes of sign in the interval between a and $a + t$).

The length of the time interval in question equals $|t|$ whether $a + t$ precedes or follows a. Since the probability of a change of sign in an elementary time interval is independent of what happens outside that interval it follows that the same is true of any interval, irrespective of where it starts. Hence the probabilities which enter into $C(t)$ are independent of the starting point a of an interval of length t, and may be derived from equation (4.35) by putting $t = t_0$. Assuming $t_0 > 0$

$$\overline{x(a) \times (a + t_0)} = \mu^2(f(0) + f(2) + f(4) + \cdots)$$
$$- \mu^2(f(1) + f(3) + f(5) + \cdots)$$
$$= \mu^2 e^{-wt_0}\left(1 - \frac{wt_0}{1!} + \frac{(wt_0)^2}{2!} - \cdots\right)$$
$$= \mu^2 e^{-2wt_0}. \qquad (4.36)$$

Since the events in question are random we may interchange past and future so that we may put $|t|$ in place of t_0, where $|t|$ is the distance in time from

4. FLUCTUATIONS AND DIELECTRIC RELAXATION

an arbitrary time a at which time the function x had a given value $x(a)$. Equation (4.37) is of the form (4.29), where now $Q = \mu^2$ and the regression time τ is identified as

$$\tau = \frac{1}{2w}. \qquad (4.37)$$

This time is equal to the relaxation time derived for the bistable model by way of a consideration of a large assembly of N bistable dipoles (equation 2.45). The regression time in equation (2.37) was derived for a single bistable dipole using probabilities in a way which corresponds to an observation of one model dipole over a long period of time.

The regression time for the bistable model signifies half the average time which the charge spends in one well without jumping out. The randomness of the process implies that the time of stay in one well may occasionally be very much longer, or very much shorter, than 2τ.

The function $C(t)$ for the bistable model is a special case of an auto-correlation function. All such functions equal unity for $t = 0$, by definition, and equal zero for $t \to \infty$. This means that no predictions can be made for the infinitely distant future or past. For finite times the shape of the auto-correlation function depends on the process in question. For a Markoff process $C(t)$, given by equation (4.29), is always positive. However, negative values are possible in other types of stochastic processes[11]. For instance, for a process where a fluctuation is superposed on to a harmonic oscillation $C(t)$ drops at first rapidly, goes through zero and then oscillates. An initial drop cannot necessarily be interpreted in terms of a regression time alone.

Given the auto-correlation function $C(t)$ we may calculate the power spectrum corresponding to it. It can be shown[3,9] that substitution of equation (4.29) into (4.28) leads to

$$G(f) = \frac{4\tau Q}{1 + (2\pi f \tau)^2}. \qquad (4.38)$$

Equation (4.38) is similar in form to the Debye equation (2.21), apart from an additive constant. However, there is a subtle difficulty involved in equating $2\pi f = \omega$ as between the Debye equation and the power spectrum. The former refers to a frequency, or time scale, which is imposed on the dielectric by an external oscillator. The latter refers to an analysis of the spontaneous fluctuations in terms of a spectrum of frequencies. When we equate frequencies in the two cases, we compare the spontaneous noise within a small frequency interval about ω with the dielectric response to an imposed alternating field of frequency ω. The two cases may be interrelated by way of the total average power of the random process. This will be derived in two ways, from $C(t)$ and $G(f)$ so as to confirm the validity of equation (4.38). When equation (4.38) is integrated then the integral over the whole frequency interval gives, according to equation (4.24), the total average power of the random process. If the fluctuating variable in question is P, we then have

$$\overline{(P - \bar{P})^2} = \int_0^\infty G(f)\,df$$

$$= 4\tau Q \int_0^\infty \frac{df}{1 + (2\pi f\tau)^2}$$

$$= Q. \tag{4.39}$$

The same value follows from (4.29), when Q is defined as the total mean square fluctuation of the polarization.

However, the fluctuation of P for the bistable model may be expressed in terms of the dielectric constant according to equation (4.18) when

$$\overline{(P - \bar{P})^2} = \frac{1}{v^2}\frac{kTv}{4\pi}(\varepsilon_s - 1)$$

$$= \frac{kT}{4\pi v}(\varepsilon_s - 1) \tag{4.18a}$$

and this equals Q according to equation (4.39). When this value for Q is introduced into equation (4.38) it becomes

$$G(\omega) = \frac{4\tau kT}{4\pi v}\cdot\frac{(\varepsilon_s - 1)}{1 + \omega^2\tau^2} \tag{4.38a}$$

while the Debye equation for $\varepsilon'(\omega)$ may be written, for $\varepsilon_\infty = 1$

$$\varepsilon'(\omega) - 1 = \frac{\varepsilon_s - 1}{1 + \omega^2\tau^2}. \tag{2.21a}$$

Comparison of the two expressions shows that

$$G(\omega) = \frac{\tau kT}{\pi v}(\varepsilon'(\omega) - 1) \tag{4.40}$$

for the simplest form of the bistable model. Now ω defines the frequency at which the noise characteristic for the spontaneous fluctuations of the model is measured. We are here not looking at a single dipole, as in the derivation of (4.38), but on a dielectric containing many bistable dipoles. For a dielectric specimen of given ε_s the number of dipoles is proportional to the volume. In view of the definition of τ by equation (4.37) the noise is inversely proportional to w.v. That is, the noise is smaller the more jumps occur in the specimen/unit time.

E. THE PHYSICAL SIGNIFICANCE OF THE RELAXATION TIME AND THE THEORY OF RATE PROCESSES

The derivation of the auto-correlation time for the bistable model relates τ to the number of jumps/unit time. This number is, according to equation (4.34), given by

$$w = N_0 \frac{p}{t_0} = w_0 p \tag{4.41}$$

where w_0 is the number of tries/unit time.

The significance of the number of tries does not follow from the properties of the model as these have been defined so far. It would be possible to elaborate the bistable model further, by assuming, for instance, that a charge in a well performs harmonic oscillations. One would then further have to assume that a large number of bistable dipoles form an assembly and that they exchange energy with each other, that a temperature T can be defined[1] for the assembly as a whole, so that each degree of freedom has an average energy $\frac{1}{2}kT$. Each oscillator would then have an average energy kT, but its energy would fluctuate. With such a model, one might attempt to describe w_0 as the the number of oscillations/unit time, while p would be the probability that a charge acquires the energy W needed to jump the potential hill between the two wells.

There is little to be gained here by elaborating the dynamic properties of the bistable model since this would require some further information on how dipoles interchange energy, and the merit of the model lies in its simplicity. The quantity w is an absolute reaction rate, and the calculation of absolute reaction rates for real physical systems is a subject in itself. In the context of dielectric relaxation this subject is known as the theory of rate processes, and a text book[12] with this title is much used. Absolute reaction rates in liquid solutions[13] and in gases[14] have recently been reviewed.

The earliest attempt to calculate the rate constant of a chemical reaction from first principles refers to the gas reaction

$$H + H_2 \rightleftharpoons H_2 + H$$

where the molecule on the left side is para-, that on the right side ortho-hydrogen, that is the nuclear spins in the left-hand molecule are anti-parallel, those on the right side parallel. The reaction has been considered[13] with the simplifying assumption that all three atoms collide along one straight line.

For the reaction in question the number of attempts may be considered as the number of collisions while the success of an attempt depends on whether the available kinetic and vibrational energy suffices to overcome the repulsion forces between the atom and molecule in collision. The calculations[14] operate with a potential hill as in the bistable model. However, this hill is three dimensional in that the potential energy is a function of two variables and the hill has a complicated structure as if a mountaineer had to traverse a winding high pass between two mountains[15]. There are different ways of crossing, some more direct than others, some involving a higher ascent than others. The probability of a successful conversion from an ortho- to a para-molecule depends not only on the minimum level of potential energy that has to be reached but also on the configuration of the course.

The case treated is particularly simple since only hydrogen is involved and the collisions occur along a straight line. Even so the calculations give only approximate results for the rate of conversion from ortho- to para-hydrogen.

The theory of rate processes operates with rate constants rather than relaxation times. For a reaction $A \to B$ a rate constant K may be defined by the equation

$$\frac{dN_A}{dt} = KN_A. \tag{4.42}$$

Where N_A is the number of molecules of species A and K is the number of molecules A disintegrating/unit time. This equation is a relaxation equation. The definition chosen implies that the end product B is removed so that the speed of the reaction depends entirely on the rate at which species A decomposes, as in radioactive decay. Equation (4.42) implies a relaxation time which will be described as

$$\tau_{equ} = \frac{1}{K}. \tag{4.43}$$

In terms of the bistable model one may identify K with w. However, the dielectric relaxation time refers to an experimental situation where the "end product" B is retained and is present in a quantity approximately equal to that of species A. Comparison of equation (4.43) with equations (2.45) or (4.37) for the dielectric relaxation time shows that

$$2\tau_{diel} = \tau_{equ}. \tag{4.44}$$

Other rate constants may be defined for other reactions and situations.

The theory of rate processes operates with the concept of a potential hill in the sense that for the reaction $A \to B$ to proceed species A has to obtain energy in order to be able to change into species B. It has to attain an "activated" state which may be pictured as somehow located at the top of a potential hill. Activation involves a change of potential energy and may also involve a change in entropy. A free energy of activation may be defined as

$$\Delta G^{++} = \Delta H^{++} - T\Delta S^{++} \tag{4.45}$$

4. FLUCTUATIONS AND DIELECTRIC RELAXATION

where the physical meaning of the energy and entropy term need not be too closely scrutinized. For the bistable model in the absence of an external field the free energy of activation would equal W unless some other terms followed from a more elaborate model which specified the dynamic properties in terms of collisions with entities other than the bistable dipoles.

The reaction rate $K = w$ may be defined in terms of the free energy of activation as

$$K = \chi \frac{kT}{h} e^{-\frac{\Delta G^{++}}{kT}} \qquad (4.46)$$

where h is Planck's quantum. χ is the "transmission coefficient" which takes into account the possibility that a specimen of species A may obtain an energy sufficient to allow it to reach the top of the potential hill but that it may not make use of it.

The term kT/h is of the dimension of a reciprocal time, and $hv = kT$ defines the average frequency of the oscillations of a harmonic oscillator in an assembly of temperature T. Around room temperature this frequency is of the order 10^{13} sec^{-1}.

The dielectric relaxation time stands in a reciprocal relationship to K and w, as shown by equation (4.43) and elsewhere. The relationship between τ and T always contains an exponential term but differs in detail according to the physical model envisaged. For the bistable model it is simplest to assume a temperature independent number of "tries"/unit time (see equation 4.41 and subsequent discussion). This leads to an equation

$$\tau = A \, e^{\frac{W}{kT}} \qquad (4.47)$$

where A and W are constants. It will be seen that this expression fits experimental data in many cases, particularly for ionic solids (see Chapter 12).

Where a rate process is assumed, another expression

$$\tau = \frac{B}{T} e^{\frac{W}{kT}} \qquad (4.48)$$

is an appropriate approximation, B and W being constants. This equation is used successfully in many cases, particularly for polymers (Chapter 13).

The pre-exponential factor A or B/T is typically of the order 10^{-13} sec^{-1} at room temperature but values differing by three or four orders of magnitude are not uncommon. Doubts about relaxation by a process which involves thermally activated movements of atoms, ions or molecules arise only when the pre-exponential factor is abnormal by many orders of magnitude, as is the case with some relaxation processes at low temperatures (see Chapter 14). Even there the anomaly of this factor does not by itself lead to firm conclusions. The magnitude of W is more significant. This energy ranges normally between some hundredths of an electron volt and a few electron volts or, say, 1 kcal/mole and 100 kcal/mole. When W is so small that $W \ll kT$ no longer holds, there is reason to doubt the applicability of classical statistical thermodynamics, that is the processes in question may involve quantum mechanical effects (see Chapters 14 and 16). However, this is exceptional.

REFERENCES

1. R. H. Fowler and E. A. Guggenheim (1939). "Statistical Thermodynamics." Cambridge University Press, London.
2. R. C. Tolman (1938). "The Principles of Statistical Mechanics." Oxford University Press, London.
3. C. Kittel (1958). "Elementary Statistical Physics." Wiley, New York.
3a. R. Kubo (1966). *Rep. Prog. Phys.*, **29**, 255.
4. M. Lax (1960). *Rev. Mod. Phys.*, **32**, 25.
5. R. E. Burgess (1958). *Can. J. Phys.*, **36**, 1569.
6. H. Fröhlich (1958). "Theory of Dielectrics." Oxford University Press, London.
7. B. K. P. Scaife (1964). "Progress in Dielectrics." 5th Edition. Birks & Hart, Heywood & Company.
8. R. E. Burgess (editor) (1965). "Fluctuation Phenomena in Solids." Academic Press, New York and London.
9. J. Bendat (1958). "Principles and Applications of Random Noise Theory", p. 213, 215. Wiley, New York.
10. S. O. Rice (1954). "Selected Papers on Noise and Stochastic Processes", pp. 133–183, p. 176. (N. Wax, ed.) Dover Publications, London.
11. R. Fürth (1954). The Physical Society (London). Report of 1954 Conference on "The Physics of the Ionosphere", p. 140.
12. S. Glasstone, K. J. Laidler and H. Eyring (1941). "The Theory of Rate Processes." McGraw Hill, New York.
13. H. Eyring and D. W. Urry (1963). *Ber. Bunsengesellschaft*, **67**, 731. (In English).
14. V. N. Kondrat'ev (1964). "Chemical Kinetics of Gas Reactions." Pergamon Press, London.
15. H. Pelzer and E. Wigner (1932). *Z. Phys. Chem.*, B, **15**, 445.

CHAPTER 5

Systems with many Relaxation Times

A. THE EXPERIMENTAL EVIDENCE

In the previous chapters it was shown that the essential physical characteristics of dielectric relaxation can be expressed in terms of a first-order linear differential equation with a constant relaxation time τ. This result follows from several lines of reasoning. The basic relaxation equation leads to the Debye equations for the complex dielectric constant, expressed in equations (2.21) and (2.22) and Fig. 2 of Chapter 2. On the basis of the theory given it would seem at first sight that most dielectrics ought to show Debye behaviour. However, this is not so: most dielectrics give measured values for $\varepsilon'(\omega)$ and $\varepsilon''(\omega)$ which deviate very widely from Fig. 2 of Chapter 2.

In summarizing the observed data one has to bear in mind that dielectric measurements usually have a practical motivation, and that insulating materials often are more or less complicated mixtures. However, a wealth of data is available for pure substances also. In summarizing the literature three classes of characteristics may be distinguished.

1. Dielectrics with a single relaxation time. These are rare; they are found among certain dilute solutions of large polar molecules in non-polar solvents, where Debye's idealized treatment (Chapter 2, Section B) may be expected to apply (see Chapter 11, Section A). Besides, some other liquids and solids obey the Debye equation, for structural reasons which tend to be complicated.
2. Dielectrics with approximate Debye behaviour, that is with peaks of ε'' as a function of $\log \omega$ which resemble Debye peaks, even though they are wider and have a somewhat different shape. This is a large class, and contains many simple compounds and solutions of dipolar compounds in non-polar solvents.
3. Dielectrics for which Debye peaks are either absent or so wide and/or deformed as to be unrecognizable. This class contains many practical insulating materials as well as some pure compounds. Classes 2 and 3 obviously cannot be rigidly separated.

The complicated and varied frequency dependence of the observed dielectric loss was historically an obstacle to the understanding of dielectric relaxation.

However, this variation does not contradict the theoretical argument. This argument only implies that the polarization of a given dipole, or polarizable element, should obey a first order linear differential equation as a function of the electric field acting on it. It does not imply that all dipoles in a given dielectric should have the same relaxation time.

B. THE SUPERPOSITION OF DELAYED RESPONSES; DECAY FUNCTIONS

The experimental evidence on dielectric relaxation can, with few exceptions, be described by assuming that a dielectric material contains a number of polarizable elements and that each of these obeys a linear differential equation

$$\tau_K \frac{dP_K}{dt} + P_K = \alpha_K E(t) \tag{5.1}$$

where τ_K and α_K are characteristic for the polarizable element in question and E is the external field which is impressed on the system. The number K may be so large as to represent a continuous gradation.

The definition of elements K in equation (5.1) is not straightforward. We have to bear in mind a complication which will be more fully discussed in Chapter 8. Electrostatic interaction between dipoles has three effects. Firstly, it introduces additional terms into the equation (5.1). This complication is very serious in the case of ferroelectrics (see Chapter 16). In general, it is relatively unimportant, insofar as electrostatic interaction only causes a not large change in the effective numerical values of α_K and τ_K for a given dipole. Besides electrostatic interaction may (or may not) have the effect that a dipole, responding with a single relaxation time in the absence of electrostatic interaction, has to be described by several terms K in the presence of electrostatic interaction.

In normal dielectrics (defined below) electrostatic interaction does not invalidate the formulation of a set of equations (5.1) for a given dielectric. It merely implies that the numerical values for α_K and τ_K cannot be simply ascribed to identifiable structural entitles, such as a given dipolar molecule.

Equations (5.1) are a set of first order differential equations with constant coefficients (τ_K and α_K), and therefore linear differential equations. Second order differential equations are also linear as long as they have constant coefficients, as is the case for LCR circuits. Linear differential equations in general have some very distinctive features. The most obvious of these is the proportionality of stimulus and response. If the equation holds for y it also holds for Ay, since (taking the example of a second order linear differential equation)

$$\omega_0^2 \frac{d^2(Ay)}{dt^2} + \tau \frac{d(Ay)}{dt} + Ay = Ax \tag{5.2}$$

for any constant A. In terms of dielectrics this means that the dielectric constant is independent of voltage. This feature is found in most dielectrics for fields up to 10^5–10^6 V cm^{-1}, and constitutes a criterion for "normal" dielectrics. Non-linear effects with regard to voltage or field will be discussed in Chapters 14–16 of this monograph.

Where the response to a stimulus is instantaneous the linear relationship between the two variables can be expressed very simply, as in the case of the optical polarization. If at some time a field E_1 is applied to the optically polarizable dielectric and if at some other time the field is changed by an increment ΔE, the optical polarization due to the combination of these two stimuli is

$$P_\infty(E_1 + \Delta E) = \alpha_\infty E_1 + \alpha_\infty \Delta E \tag{5.3}$$

quite generally. The effects of the two stimuli add without influencing each other. The dipolar polarization described by equations (5.1) responds to stimulation by an applied field with a delay, and delayed responses have to be added.

We shall in the first place consider a single element of type K which obeys equation (5.1) with the single time constant τ_K and assume that the dielectric is subject to a field $E(t)$ as in Fig. 1(a). This field is zero from $t = -\infty$ to t_1 when it changes in one step to E_1. At t_2 a second step of magnitude $-\Delta E$ occurs, reducing the field to E_2. This may be considered as two stimuli, at t_1 and t_2, the responses to which add without influencing each other. The

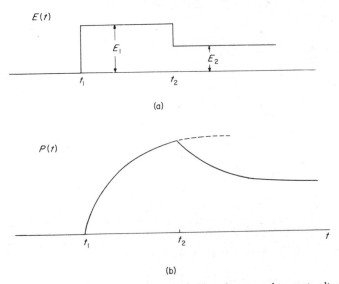

FIG. 1. Field and polarization for a linear dielectric subjected to two subsequent voltage steps. $E = 0$ for $t < 0$, $E = E_1$ for $t_1 < t < t_2$ and $E = E_2$ for $t > t_2$.

responses to the two steps E_1 and $-\Delta E$ can be calculated as in equation (1.11), and their addition gives for a time $t > t_2$

$$P(t) = E_1 \alpha_K (1 - e^{-\frac{t-t_1}{\tau_K}}) - (E_1 - E_2)\alpha_K (1 - e^{-\frac{t-t_2}{\tau_K}}). \quad (5.4a)$$

The superposition of the after-effects of the two stimuli give $P(t)$ as sketched in Fig. 1(b).

Equation (5.4a) can be expressed as the sum of two integrals

$$P(t) = E_1 \int_{t_1}^{t_2} \frac{\alpha_K}{\tau_K} e^{-\frac{t-u}{\tau_K}} du + E_2 \int_{t_2}^{t} \frac{\alpha_K}{\tau_K} e^{-\frac{t-u}{\tau_K}} du \quad (5.4b)$$

where E_1 is the field operating between t_1 and t_2 and E_2 the field thereafter. More generally $P(t)$ can be expressed by an integral expression

$$P(t) = \int_{-\infty}^{t} E(u)\dot\Psi(t-u) \, du \quad (5.4c)$$

of which equation (5.4b) is a special case. The more general expression cannot be rigorously derived by the elementary use of step-functions, as done here. A more general discussion is given in Appendix IV.

The function $\dot\Psi(t-u)$ in equation (5.4c) may be written $\dot\Psi(t)$ where now t denotes a time interval. For a single element equation (5.4b) shows that

$$\dot\Psi(t) = \frac{\alpha_K}{\tau_K} e^{-\frac{t}{\tau_K}}. \quad (5.5a)$$

If there are n_K elements with characteristics K their contribution to P is given by $n_K \Psi_K$. For the dielectric as a whole

$$\dot\Psi(t) = \sum_K \frac{n_K \alpha_K}{\tau_K} e^{-\frac{t}{\tau_K}} \quad (5.5b)$$

since all contributions superpose linearly.

The function $E_0 \dot\Psi(t)$ is of the dimension of a polarization divided by a time and hence of a current density. This function decays exponentially with time and may thus be called a decay function. There is some confusion here in the literature, in that $F_0 \dot\Psi(t)$ describes the decay of a quantity of dimension of a current density while $P(t)$ itself can also be described by a decay function. This may be seen if we envisage a step-function such that $E = E_0$ from $t = -\infty$ to $t = 0$ and $E = 0$ for $t > 0$. That is, if we discharge a fully charged dielectric. Now integration of equation (5.1) for a single element K gives

$$P(t) = E_0 \alpha_K e^{-\frac{t}{\tau_K}} = E_0 \Psi(t) \quad (5.6a)$$

5. SYSTEMS WITH MANY RELAXATION TIMES

where $\Psi(t)$ is the decay function of the polarization itself. For the dielectric as a whole

$$\Psi_P(t) = \sum_K n_K \alpha_K \, e^{-\frac{t}{\tau_K}} \tag{5.6b}$$

which apart from the sign is an integral of the function in (5.5b). The two functions $\Psi(t)$ and $\dot{\Psi}(t)$ have the same form but different physical dimensions. We shall call them potential decay function and current decay function respectively.

A decay function for a single relaxation time of an element obeying a first order linear differential equation (5.1) is given by a constant multiplied by an exponential term, as in (5.5a) and (5.6a). This is the same form as equation (4.29) which has been derived for the bistable model as an auto-correlation function, while the random nature of thermal motion was exactly defined. Decay functions may be understood in terms of fluctuations where adequate structural data are available. For systems obeying differential equations with second order terms as equation (5.2) the basic form of the decay function is no longer a simple exponential dependence on time, as will be mentioned in Chapter 9.

C. INTEGRAL FORMULATIONS OF THE PRINCIPLE OF SUPERPOSITION, AND THE DISTRIBUTION FUNCTION

So far decay functions have been described for stimuli by simple step-functions, while an integral formulation of the superposition of delayed responses was quoted as equation (5.4c). The principle of superposition is normally treated in terms of integral expressions[1,2,3,4]. Such a formulation, if rigorous, demands mathematical methods which go beyond the scope of this monograph (see 2, 4 and Appendix IV). The integral expressions of the principle of superposition are apt to be confusing because the same statement can be made in many alternative ways. Appendix IV gives different transformations of equation (5.4c) while one of these transformations appears as equation (6.5).

The generalization of equations (5.5b) and (5.6b) to an integral expression is straightforward for the case of step-function stimuli as envisaged here. For this purpose the polarizable elements are divided into groups of given relaxation time. The number of dipoles whose relaxation time lies between τ and $\Delta \tau$ may be defined as

$$\Delta n(\tau) = f(\tau) \, \Delta \tau \tag{5.7}$$

while the polarizability of these elements is $\alpha(\tau)$. The static polarization contributed by these elements is for a constant field, which has operated since $t = -\infty$,

$$\Delta P_s = E_0 \alpha(\tau) f(\tau) \Delta \tau. \tag{5.8}$$

In view of the definition of the static dielectric constant (equation 2.10) this implies

$$\Delta \varepsilon_s = 4\pi \alpha(\tau) f(\tau) \Delta \tau \tag{5.9}$$

and we define a distribution function by

$$y(\tau) = \frac{d\varepsilon_s(\tau)}{d\tau} = 4\pi \alpha(\tau) f(\tau) \tag{5.10}$$

where

$$\int_0^\infty y(\tau) \, d\tau = \varepsilon_s - \varepsilon_\infty. \tag{5.11}$$

Equation (5.6b) for the decay function of the polarization may now be formulated for the continuous distribution defined. This gives

$$\Psi_P(t) = \frac{1}{4\pi} \int_0^\infty y(\tau) e^{-\frac{t}{\tau}} \, d\tau. \tag{5.6c}$$

The integral formulation for the current decay function may be deduced by generalization of equation (5.5b) as

$$\dot{\Psi}_P(t) = \frac{1}{4\pi} \int_0^\infty y(\tau) \frac{1}{\tau} e^{-\frac{t}{\tau}} \, d\tau. \tag{5.5c}$$

The current decay function multiplied by a field determines a time derivative of $P(t)$ for a step-function, that is a charging or discharging current, as will be discussed in more detail in Section A of Chapter 6. (For the discharge described by equation (5.6a) the current would be negative, while for the purposes of equation (5.4c) the step "down" in field would count as negative.)

Given an experimentally determined decay function the distribution function can be determined. As a first step to this determination it is convenient to introduce a new variable $z = 1/\tau$ and a new distribution function

$$N(z) = \frac{1}{z^2} y\left(\frac{1}{z}\right). \tag{5.12}$$

Now the decay functions may be written in the form

$$\Psi_P(t) = \frac{1}{4\pi} \int_0^\infty N(z) e^{-tz} \, dz \tag{5.6d}$$

5. SYSTEMS WITH MANY RELAXATION TIMES

for the potential decay function and

$$\dot{\Psi}_P(t) = \frac{1}{4\pi} \int_0^\infty N(z) z \, e^{-tz} \, dz \tag{5.5d}$$

for the current decay function. These two expressions are Laplace integrals and $N(z)$ can be derived from the decay functions by standard methods for the inversion of Laplace integrals. References 2–4 quote the literature of operational calculus of which this procedure is part and Appendix IV contains a few useful formulae and references.

The integral formulation of decay functions is straightforward where $E(t)$ is simply defined as for a step-function or a sinusoidal field. When an integral form of a decay function is introduced in a more general way into an integral expression of superposition as in equation (5.4c), mathematical complexities arise which are the subject of a considerable literature[2,4]. Some of the complexities are due to a consideration regarding the validity of equation (5.1) which will be discussed in Section E while others are due to the circumstance that $E(t)$ may be a function which is inconvenient for the formulation of an integral. Besides, a variety of notations are currently used. None of these complications are relevant for normally conducted experiments on dielectric relaxation. However, they sometimes come into play when dielectric and mechanical responses are compared, the information being sufficiently detailed to make the form of the distribution function important. Relevant literature will be quoted in Chapter 13, which is concerned with polymers.

All three equations (5.4) are expressions of the principle of superposition, (5.4c) being the most general, while Appendix IV treats a further generalization. The principle of superposition was verified in a very sophisticated way by Hopkinson[1] in 1876. Hopkinson measured the charge on a Leyden jar to which step-functions in voltage were applied at discrete times, and analysed the results in terms of an integral as in (5.4c). Making allowance for the direct conductivity of glass and capacitative and inductive surges he found the principle of superposition confirmed. Later workers have come to the same conclusion, in so far as dielectric relaxation can be treated in terms of equations (5.1).

The validity of the principle of superposition proves that dielectric relaxation obeys linear differential equations. That is, in terms of stimulus s and response r it proves that terms s^2, $s\frac{ds}{dt}$, r^2, $r\frac{ds}{dt}$, rs and other quadratic or otherwise non-linear terms are absent from the differential equations which describe the time dependence of P and D. This principle does not prove that higher order differentials such as $\frac{d^2s}{dt^2}$ or $\frac{d^2r}{dt^2}$ are absent. Equation (5.2) is an equation for a resonant system and may apply to dielectric relaxation, though only at very high frequencies as discussed in Chapter 9. In the vast majority of cases the term containing $\frac{d^2P}{dt^2}$ is quite negligible and the response of the dielectric can be represented by a system of equations (5.1).

D. THE COMPLEX DIELECTRIC CONSTANT

The decay functions for a system obeying a set of differential equations (5.1) are defined with respect to an applied field $E(t)$ which is envisaged as a series of steps. Equation (5.4c) is applicable also to a sinusoidal applied field, as will be discussed in more detail in Section A of Chapter 6.

A group of polarizable elements with relaxation times in the frequency interval τ to $\tau + \Delta\tau$ contributes a term $\Delta\varepsilon_s$ to the dielectric constant according to equation (5.9). Such a group has a single relaxation time and thus contributes a Debye expression to the complex dielectric constant.

The total dielectric constant of the material is given by the sum of the contributions of all groups. This sum can be written as an integral, for a continuous distribution as defined by equation (5.10). Separating real and imaginary parts,

$$\varepsilon'(\omega) - \varepsilon_\infty = \int_0^\infty \frac{y(\tau)\, d\tau}{1 + \omega^2\tau^2} \tag{5.13}$$

$$\varepsilon''(\omega) = \int_0^\infty \frac{y(\tau)\omega\tau}{1 + \omega^2\tau^2}\, d\tau. \tag{5.14}$$

These equations replace the Debye equations (2.21) and (2.22) which hold if $y(\tau) = 4\pi N\alpha$ for $\tau = \tau_0$ and is zero elsewhere.

When $\varepsilon^*(\omega)$ for a material is known the distribution function may be deduced from the data. This is obvious, in principle, from the inspection of equations (5.13) and (5.14), but the mathematical tools required are relatively advanced (see Appendix V). However, an important relationship may be deduced quite simply.

Equations (5.13) and (5.14) imply that there exists a relationship between the polar static dielectric constant $\varepsilon_s - \varepsilon_\infty$ and the area under the curve plotted for ε'' as a function of $\log \omega$. This can be proved by using the identity

$$\int_0^\infty \frac{\omega\tau}{1 - \omega^2\tau^2}\, d\log_e \omega = \frac{\pi}{2} \tag{5.15}$$

in evaluating the area under the loss curve, namely

$$\int_0^\infty \varepsilon''(\omega)\, d\log_e \omega = \int_{\tau=0}^\infty \int_{\omega=0}^\infty \frac{y(\tau)\omega\tau}{1 + \omega^2\tau^2}\, d\tau\, d\log \omega.$$

5. SYSTEMS WITH MANY RELAXATION TIMES

The calculation gives

$$\int_0^\infty \varepsilon''(\omega) \, d\log_e \omega = \frac{\pi}{2} \int_0^\infty y(\tau) \, d\tau \tag{5.16}$$

while the integral over $y(\tau)$ can be expressed also by equation (5.11) so that

$$\int_0^\infty \varepsilon''(\omega) \, d\log_e \omega = \frac{\pi}{2} (\varepsilon_s - \varepsilon_\infty) \tag{5.17}$$

an important result which is a special case of the Kramers–Kronig relations discussed in Section E of this chapter.

Equation (5.17) can be verified for dielectric of groups 1 and 2, wherever $\varepsilon''(\omega)$ has been measured over a wide range and can be assumed as zero outside the range of frequencies covered.

The frequency range available can be stretched out very considerably by using measurements at different temperatures. This method uses equation (4.47) or (4.48) which implies that the position of the Debye peak ($\omega\tau = 1$) shifts with temperature according to

$$\omega_{\text{peak}} = \frac{T}{B(T)} e^{\frac{W}{kT}}. \tag{5.18}$$

Incidentally, equation (5.17) can be used to determine $\varepsilon_s - \varepsilon_\infty$ for weakly polar materials, where it may be easier to measure $\varepsilon''(\omega)$ over a wide frequency than to determine the static dielectric constant accurately. Sillars[5] used this method in 1938, in conjunction with measurements at different temperatures.

E. AN AMBIGUITY CONCERNING THE DISTRIBUTION FUNCTION: RETARDATION AND RELAXATION

The derivation of a distribution function $y(\tau)$ from measured data of $\varepsilon^*(\omega)$ or responses to step-functions in field is analogous to the problem of analysing two terminal measurements carried out on a black box, containing an unknown combination of capacitors and resistors. The derivation of $y(\tau)$ signifies that an equivalent circuit is required in terms of parallel branches, each containing a capacitance and resistance in series.

The derivation of distribution functions from the response of a material to step-function stimuli formulated in a general way raises a problem which is lucidly discussed by Gross[2]. The problem can be stated for the dielectric case with the help of the equivalent circuit sketched in Fig. 2. The figure also indicates the procedure whereby a step-function in voltage is applied.

The network in Fig. 2 is equivalent to a dielectric with $\varepsilon_\infty = 0$ and without direct conductance. Each individual element obeys the differential equation

(1.7), which is analogous to an equation for element K in equations (5.1), with a charge Q_K in the place of the polarization P_K.

Figure 2 specifies a procedure as well as a circuit. It is assumed that the battery of voltage V_0 has been charging the composite capacitor for an

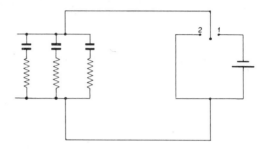

FIG. 2. Equivalent circuit for a capacitor which contains a dielectric with a distribution of relaxation times. For the discharge envisaged first the switch has been in position 1 from $t = -\infty$ to $t = 0$ and is switched to position 2 at $t = 0$.

infinite time, and that at $t = 0$ the battery is replaced by a short circuit. This implies that each element now discharges independently of all others, so that

$$Q_K(t) = V_0 C_K e^{-\frac{t}{\tau_K}} \qquad (5.19)$$

where $\tau_K = C_K R_K$. The total charge of the composite capacitor is the sum of the charges on the individual ones, and represents $4\pi P(t)$ for the dielectric to which the circuit is analogous.

For the circuit in question, a step-function applied as indicated in Fig. 2 leads to a certain $P(t)$ which implies a certain distribution function $y(\tau)$ which may be derived using equation (5.6d). A step-function applied in another way leads to a $P(t)$ which corresponds to a different distribution function. With the procedure indicated in Fig. 2 the battery is disconnected at $t = 0$ and replaced by a short circuit. However, the capacitors could also be discharged if the leads to the battery were simply taken away at $t = 0$. In that case, the individual elements would be partly in series, partly in parallel, and another distribution function would follow from the decay of the total charge Q on all the capacitors. Equation (5.1) with the appropriate numbers n_K of polarizable elements of type K would then no longer apply to the material in question. The appropriate distribution function would not be $y(\tau)$ defined by equation (5.1) (which in accordance with the mechanical analogue is called the "retardation" function) but another distribution function, namely the "relaxation" function.

The distinction between retardation and relaxation functions is rarely

5. SYSTEMS WITH MANY RELAXATION TIMES

relevant in the electrical case. Whenever $\varepsilon^*(\omega)$ is determined, whether by alternating current or step-function methods, the same $y(\tau)$ applies. Experiments where a fixed charge Q is applied to a material and its decay observed are aimed to elucidate the complexities of charge transport in dielectrics and will not be discussed in this monograph. However, the distinction between relaxation and retardation is important for mechanical experiments where sometimes the stress and sometimes the strain is impressed on a material. This intricate subject is treated by Gross[2]. If a distribution function has been determined for a given material and procedure the distribution function for the same material and a different procedure can be calculated.

A number of distribution functions $y(\tau)$ have been derived by numerous investigators from measured relaxation data. Gross[2] quotes some of these in his monograph, and a few will be discussed in Chapters 7 and 13. As a rule, it is easier to calculate relaxation data on the basis of a known $y(\tau)$ than to perform the inverse operation. The usual method of interpreting measurements of $\varepsilon^*(\omega)$ or $P(t)$ is to fit a likely distribution function to the data. If the relaxation data are not very detailed this fitting is usually simplified by the graphical methods described in Chapter 7.

F. THE KRAMERS–KRONIG RELATIONS

The most general statement that can be made about relaxation in linear systems concerns the relationship between the real and imaginary part of the dielectric constant. That such a relationship must exist can be deduced qualitatively from the inspection of equations (5.13) and (5.14). Both $\varepsilon'(\omega)$ and $\varepsilon''(\omega)$ are functions of the same $y(\tau)$, and $y(\tau)$ can in principle be calculated from either. This is only possible if $\varepsilon'(\omega)$ and $\varepsilon''(\omega)$ are functions of one another. The connection between $\varepsilon'(\omega)$ and $\varepsilon''(\omega)$ is given by the so-called Kramers–Kronig relations. These apply generally to all dissipative systems, and presuppose that

1. The system is linear.
2. The constitution of the system does not change within the time contemplated.
3. The system obeys the causality principle, in that the response does not precede the stimulus.

It is difficult to derive the Kramers–Kronig relations using only elementary mathematics.

Relatively simple derivations are given by Gross[2], Fröhlich[6], and by

Kittel[7], while Toll[8] discusses the Kramers–Kronig relations in a wide context. They are defined as

$$\varepsilon'(\omega) - \varepsilon_\infty = \frac{2}{\pi} \int_0^\infty \frac{u\varepsilon''(u)}{u^2 - \omega^2} du \qquad (5.20)$$

$$\varepsilon''(\omega) = -\frac{2\omega}{\pi} \int_0^\infty \frac{\varepsilon'(u) - \varepsilon_\infty}{u^2 - \omega^2} du. \qquad (5.21)$$

The variable of integration, u, is real; the principal parts of the integrals are to be taken in the event of singularities of the integrands. Equation (5.21) implies that $\varepsilon''(\infty) = 0$.

The evaluation of equations (5.20) and (5.21) is laborious if $\varepsilon^*(\omega)$ is not a convenient analytical function. However, the calculation of $\varepsilon_s - \varepsilon_\infty$ is simple, in that it reduces to the equation (5.17) which was quoted earlier.

G. AN HISTORICAL NOTE

The principle of superposition was first formulated by Boltzmann[9] in 1874, and was applied to dielectrics in 1876 by Hopkinson[1]. Later a great deal of confusion arose in connection with the significance of distribution functions. This confusion did not arise as long as only macroscopic properties were considered. Wiechert[10] first introduced distribution functions in 1893. Von Schweidler[11] in an important review article in 1907, based his treatment on a system of differential equations as done in the present monograph, without enquiring about models. Rather unfortunately, a search for models took a confusing turn in 1913, when Wagner[12] pointed out that a treatment by Maxwell[13] for a dielectric consisting of layers of different conductivity leads to a set of differential equations like (5.1), and that this treatment could be generalized to other composite dielectrics. Maxwell himself considered his treatment as no more than a mathematical exercise which showed how a dielectric with many relaxation times can arise. He did not preclude the possibility that molecular mechanisms might lead to the same result. Some workers following Wagner attributed too much importance to the Maxwell–Wagner effect which they used to explain the behaviour also of homogeneous dielectrics. This confusion persists to some extent even now in that some authors consider the mathematical deduction of a distribution function tantamount to a physical explanation.

REFERENCES

1. J. Hopkinson (1901). Original papers, pp. 1–43, 119–153. Cambridge University Press, London.
2. B. Gross (1953). "Mathematical Structure of the Theories of Viscoelasticity." Hermann et Cie, Paris.

3. J. T. Bergen (editor) (1960). "Viscoelasticity, Phenomenological Aspects." Academic Press, London and New York.
4. J. R. MacDonald and C. A. Barlow, Jr. (1963). *Rev. Mod. Phys.*, **35**, 940.
5. R. W. Sillars (1938). *Proc. R. Soc. Ass.*, **169**, 66.
6. H. Fröhlich (1958). "Theory of Dielectrics." Clarendon Press, Oxford.
7. C. Kittel (1958). "Elementary Statistical Physics." Chapman and Hall, London.
8. J. S. Toll (1956). *Phys. Rev.*, **104**, 1760.
9. L. Boltzmann (1874). *Akad. Wissenschaft, Wien.*
10. E. Wiechert (1893). *Wied. Ann. Phys.*, **50**, 335.
11. E. von Schweidler (1907). *Annalen Phys.*, (4) **24**, 711.
12. K. W. Wagner (1913). *Annalen Phys.*, (4) **40**, 817.
13. J. C. Maxwell (1954). "A. Treatise on Electricity and Magnetism." Vol. 1, p. 452. Dover Publications, New York.

CHAPTER 6

Dielectric Measurements and their Interpretation

The permittivity and dielectric loss of a material are almost always measured by introducing it into a capacitor, wave guide, or other container which forms part of an electrical circuit. This circuit is subjected to an alternating voltage or to a sudden voltage step (a step-function). In either case errors may arise if effects due to some parts of the external circuit are ascribed to the specimen.

The present chapter gives a brief survey of methods of measurement, based mainly on existing reviews. The design of apparatus is not discussed, except for a few notes regarding new techniques. The main aim of the chapter is to point out the limits of experimental accuracy as well as some sources of error which might escape the less experienced user of ready-made equipment or the theoretician who examines published data.

The measurement techniques fall into distinct groups according to frequency. Step-function (direct current) methods will be considered first.

A. STEP-FUNCTION MEASUREMENTS

The use of charging and discharging currents for the determination of dielectric loss goes back to Hopkinson[1]. The method yields data for $\varepsilon^*(\omega)$ at very low frequencies, below 1 c/s and is increasingly used to complement data obtained by more usual alternating current methods. The treatment which leads to the simple practical formula (6.28) is given in detail because many experimenters find the mathematics difficult and because concern with the mathematical formalism obscures an important physical difficulty inherent in the step-function method.

It is in principle easy to apply to a specimen a step-function in voltage. In practice, the measurement of charging currents demands a voltmeter of very high input impedance. Early measurements were made using some form of quadrant electrometer, an instrument satisfactory in principle but rather inconvenient in use. More recently valve voltmeters and vibrating reed electrometers have been developed which have input impedances of up to 10^{16} Ω and which use electronic methods of amplifying small voltages so that direct current methods have become more convenient.

Measurements of charging or discharging currents are designed to yield the decay function $\dot{\Psi}(t)$ defined in Section B of Chapter 5. It will be shown that $\varepsilon''(\omega)$ can be deduced from direct current data in fair approximation, without lengthy calculations.

The mathematical treatment operates in terms of the dielectric displacement D and its derivative

$$J(t) = \frac{dD(t)}{dt} \tag{6.1}$$

which is a current density. According to Section A of Chapter 2, $D(t)$ for a dielectric is given by

$$D(t) = \varepsilon_\infty E(t) + 4\pi P_D(t)$$

$$= 4\pi(P_\infty(t) + P_D(t)). \tag{6.2}$$

The discussion in Chapter 5 refers to the "dipolar" part P_D of the polarization, that is to the polarization which relaxes with finite relaxation time. The argument of Chapter 5 will be applied so that a current decay function is defined for D rather than P when

$$\dot{\Psi}(t) = 4\pi \dot{\Psi}_P(t) \tag{6.3a}$$

in general and

$$\dot{\Psi}(t) = \frac{(\varepsilon_s - \varepsilon_\infty)}{\tau} e^{-\frac{t}{\tau}} \tag{6.3b}$$

for a single relaxation time in accordance with equations (5.5c) and (5.11).

When the decay function of P_D is used to calculate $D(t)$ the optical part of the polarization P_∞ has to be included as being equal to ε_∞ times the field. Now, using equations (5.4c) and (6.3a)

$$D(t) = \varepsilon_\infty E(t) + \int_{-\infty}^{t} E(u)\dot{\Psi}(t-u)\, du. \tag{6.4}$$

Equation (6.4) follows more elegantly from the treatment in Appendix IV (equation B).

It is convenient to introduce a variable $x = t - u$ in equation (6.4) so that

$$D(t) = \varepsilon_\infty(t) + \int_0^\infty E(t-x)\dot{\Psi}(x)\, dx \tag{6.5}$$

which is equivalent to equation (D) of Appendix IV.

It may be helpful to show how equation (6.5) is applied to step-functions.

6. DIELECTRIC MEASUREMENTS AND THEIR INTERPRETATION

For charging $E(t - x) = 0$ for $x > t$ and equals E_0 for $x < t$. Hence, for a single relaxation time

$$D(t) = \varepsilon_\infty E_0 + E_0 \int_0^t \frac{\varepsilon_s - \varepsilon_\infty}{\tau} e^{-\frac{x}{\tau}} dx$$

$$= \varepsilon_\infty E_0 + E_0(\varepsilon_s - \varepsilon_\infty)(1 - e^{-\frac{t}{\tau}}) \qquad (6.6a)$$

as expected. For a discharge $E(t - x) = E_0$ for $x > t$ and equals zero for $x < t$. Now, since $\varepsilon_\infty E(t)$ vanishes

$$D(t) = E_0 \int_t^\infty \frac{\varepsilon_s - \varepsilon_\infty}{\tau} e^{-\frac{x}{\tau}} dx$$

$$= E_0(\varepsilon_s - \varepsilon_\infty) e^{-\frac{t}{\tau}} \qquad (6.6b)$$

so that this equation represents both charge and discharge correctly. In either case, and for a more general decay function,

$$D(t) = \text{const} \mp E_0 \int_0^t \dot{\Psi}(x) \, dx \qquad (6.7)$$

where a negative sign corresponds to a discharge. Differentiation of equation (6.7) gives for a step-function, in view of equation (6.1),

$$J(t) = \pm E_0 \dot{\Psi}(t). \qquad (6.8)$$

The current always decays with time, although the direction of its flow is opposite for charge and discharge.

In practice there are two complications involved in the measurement of a decay function. Firstly, the switching at $t = 0$ is accompanied by a sudden change of $\varepsilon_\infty E_0$ in the electric displacement. Theoretically this should last an infinitely short time, that is the current corresponding to it should be a delta function. In practice, the external circuit attached to the measuring instrument does not respond infinitely fast, and the switching surge lasts for some time. Hence all time constants in the external circuit must be kept as short as possible.

A greater difficulty stems from the presence of conduction currents. Equations (6.1)–(6.6) refer to a dielectric, so that the current in equation (6.8) is defined as a displacement current. However, all materials conduct electricity, if only to an extremely slight extent. Hence, the current density is

$$J(t) = J_D(t) + J_C \qquad (6.9)$$

where J_C is a conduction current due to electronic or ionic carriers.

The role of the conduction current is not obvious when one wishes to calculate $D(t)$ by integrating the current.

$$D(t) = \int_0^t J_D(t)\,dt \qquad (6.10)$$

for the case where

$$\text{div } J_C = 0 \qquad (6.11)$$

i.e. where the conduction current behaves according to Maxwell's equations. If the contacts between the dielectric and the electrodes are ohmic, equation (6.11) applies to the current carried by any kind of carrier independently of time. However, in practice contacts are not usually ohmic (see Chapter 14, Section C). This means that a part of the electronic or ionic carriers which contribute to conduction pile up at the electrodes and build up space charges which in turn oppose the flow of current. Hence J_C decreases with time until a value J_s is reached which corresponds to a steady state and obeys equation (6.11). In this case the current may be separated into three parts

$$J(t) = J_D(t) + J_i(t) + J_s \qquad (6.12)$$

where the second term is due to conduction in the dielectric, due to the build up of space charges at the electrodes or at the ends of some internal conducting paths which do not reach the electrodes. This term leads to a polarization so that

$$D(t) = \int_0^t (J_D(t) + J_i(t))\,dt. \qquad (6.13)$$

The time-dependent parts of the current cannot be separated into the true displacement current and the current which builds up an "interfacial" polarization as long as all effects are linear. However, the build up of space charges at the electrodes implies non-linearity and may be observed in favourable circumstances, as discussed in Chapter 14. Non-linear behaviour of D as a function of E invalidates the principle of superposition, so that the complications introduced by non-ohmic electrodes and interfacial polarization are serious in principle. However, these difficulties can often be neglected for normal dielectrics which include most materials used for electrical insulation.

Figure 1 shows a typical example of charging and discharging currents observed for a good dielectric. The charging current density settles down to a constant value of J_s after a time, which may however be inconveniently long. If J_s can be detected and is subtracted from the charging current the displacement currents on charge and discharge are identical except for their direction. If this is so independently of E_0 then $\dot{\Psi}(t)$ can be measured unequivocally, whether it contains a contribution from mobile carriers or not. Satisfactory measurements of $\dot{\Psi}(t)$ have been carried out for many dielectrics as discussed later in this section and Chapter 13.

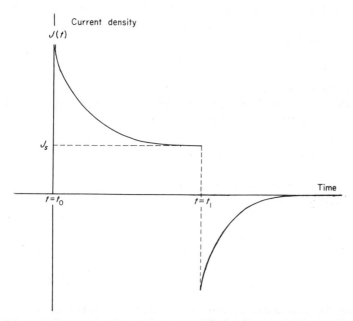

FIG. 1. Schematic diagram showing the charging and discharging current of a capacitor. A field E_0 is applied between t_0 and t_1 while $E = 0$ at all other times. $J(t)$ is the current density.

Hopkinson[1] analysed measurements of charging and discharging currents and deduced that

$$\dot{\Psi}(t) = \frac{K}{t^m} \tag{6.14}$$

is often a satisfactory expression, where K and m are constants and m is near unity.

This type of expression has been observed by many experimenters. However, it should be noted[2] that equation (6.14) cannot hold for infinitely long times, since it would give $D \to \infty$ for $t \to \infty$ if $m \leq 1$. Equation (6.14) is merely an approximate expression, which holds over considerable periods of time.

The decay function in (6.14) has been deduced from the response of a dielectric to a step-function in voltage. As outlined in Chapter 5, it implies a distribution function $y(\tau)$ which applies also to the response of the dielectric to an alternating voltage. It must hence be possible to express the step-function response in terms of $\varepsilon^*(\omega)$ by way of $y(\tau)$, using equations (5.5d) and (5.12).

In practice, it is relatively easy to express the data deduced from the response of a dielectric to a step-function in voltage in terms of the complex dielectric constant without evaluating $y(\tau)$. If the decay function of a dielectric is known its response to an alternating voltage may be deduced from equation (6.5) by putting

$$E(t - x) = E_0 \cos \omega(t - x) \tag{6.15}$$

for the sinusoidal applied field. If we neglect any after effect of the polarization at the time when the field was switched on we can retain the limits of integration of equation (6.5). Now

$$D(\omega, t) - \varepsilon_\infty E_0 \cos \omega t = E_0 \int_0^\infty \dot{\Psi}(x) \cos \omega(t - x) \, dx. \tag{6.16}$$

The dependence on time in equation (6.16) can be simplified by applying a trigonometric formula so that

$$D(\omega, t) - \varepsilon_\infty E_0 \cos \omega t = E_0 \cos \omega t \int_0^\infty \dot{\Psi}(x) \cos \omega x \, dx$$

$$+ E_0 \sin \omega t \int_0^\infty \dot{\Psi}(x) \sin \omega x \, dx \tag{6.17}$$

since t can be treated as constant with respect to the integration.

Equation (6.17) can be formulated in terms of the complex dielectric constant. The dielectric displacement may be defined as a complex quantity, in analogy to the charge in equation (1.15), so that

$$D^*(\omega, t) = \varepsilon^*(\omega) E^*(\omega, t) \tag{6.18}$$

or, for sinusoidal fields

$$D^*(\omega, t) = (\varepsilon'(\omega) - i\varepsilon''(\omega)) E_0 \, e^{i\omega t}. \tag{6.19}$$

Equating the real terms on both sides gives

$$D(\omega, t) = E_0(\varepsilon'(\omega) \cos \omega t + \varepsilon''(\omega) \sin \omega t). \tag{6.20}$$

Comparison of equations (6.17) and (6.20) shows that

$$\varepsilon'(\omega) = \varepsilon_\infty + \int_0^\infty \dot{\Psi}(x) \cos \omega x \, dx \tag{6.21}$$

$$\varepsilon''(\omega) = \int_0^\infty \dot{\Psi}(x) \sin \omega x \, dx \tag{6.22}$$

with the decay function in electrostatic units. When $\dot{\Psi}(t)$ is known, the integration of equations (6.21) and (6.22) is straightforward. It has been carried out by Hopkinson[1] for the decay function given by equation (6.14) as early as 1896, but the numerical evaluation of the integral is laborious and the subject was rather neglected for a long time. A simplified method of integration was proposed by Hamon[3].

Before discussing the approximate integration of equations (6.21) and (6.22) it is useful to clarify the role of J_s for the measured dielectric loss. J_s can be separated from the time dependent current if step-functions are used but it cannot be separated with alternating current methods. When it is desired to link up the measurements of $\varepsilon''(\omega)$ with step-function and alternating current methods allowance has to be made for the steady conductivity γ_s which corresponds to J_s.

Hamon's calculation uses practical units for the conductivity, and the dependence of $\varepsilon''(\omega)$ on a conductivity is given by the formula for the representation of a dielectric by a parallel circuit (see Appendix II)

$$\varepsilon'(\omega) - i\varepsilon''(\omega) = \frac{1}{\varepsilon_0}\left(C_p(\omega) - \frac{i\gamma(\omega)}{\omega}\right). \tag{6.23}$$

Equating the imaginary parts in equation (6.23) gives

$$\varepsilon''(\omega) = \frac{1}{\varepsilon_0 \omega} \gamma(\omega) \tag{6.24}$$

where $\gamma(\omega)$ is the parallel conductivity in the practical system chosen.

The direct conductivity γ_s is independent of frequency. Hamon[3] formulates equation (6.22) so as to include the contribution from the direct conductivity so that

$$\varepsilon''(\omega) = \frac{1}{\varepsilon_0}\left(\frac{\gamma_s}{\omega} + \int_0^\infty \dot{\Psi}(x) \sin \omega x \, dx\right) \tag{6.25}$$

while $\varepsilon'(\omega)$ is still given by equation (6.21), the decay function being given in practical units.

The integrals in equations (6.21) and (6.25) may be evaluated[1,3] if $\dot{\Psi}(t)$ is given by equation (6.14) when

$$\int_0^\infty \dot{\Psi}(t) \cos \omega t \, dt = K\omega^{m-1}\Gamma(1-m)\cos\frac{(1-m)\pi}{2} \tag{6.26}$$

$$\int_0^\infty \dot{\Psi}(t) \sin \omega t \, dt = K\omega^{m-1}\Gamma(1-m)\cos\frac{m\pi}{2} \tag{6.27}$$

where $\Gamma(1-m)$ is the gamma function. Equation (6.26) holds for $0 < m < 1$, equation (6.27) for $0 < m < 2$. Hamon shows that equation (6.27) can be greatly simplified if $0.3 < m < 1.2$ a range which covers the values of m

found in practice. He shows that $\varepsilon''(\omega)$ can be approximated by the simple formula

$$\varepsilon''(\omega) = \frac{1}{\varepsilon_0 \omega}(\gamma_s + \dot{\Psi}(t_1)) \tag{6.28}$$

where ω and t_1 are related by

$$\omega t_1 = \left(\Gamma(1-m)\cos\frac{m\pi}{2}\right)^{-\frac{1}{m}} \tag{6.29}$$

while the quantity on the right-hand side may be taken as constant so that, approximately,

$$\omega t_1 = 0.63. \tag{6.30}$$

Equation (6.28) is used as follows: the conductivity is determined as a function of time. A given time t_1 corresponds to a given ω according to equation (6.30), and $\dot{\Psi}(t_1)$ for the frequency required is entered into equation (6.28). The constant ε_0 is given in Table I of Appendix I.

Equation (6.28) might be used to calculate $\varepsilon'(\omega)$ even though equation (6.26) is valid within narrower limits than (6.27). If the superposition principle is valid the Kramers–Kronig relations follow and $\varepsilon'(\omega)$ can be calculated from equations (5.20) or (5.17). However, in this way the linearity of all effects has been assumed twice over, so that caution is indicated. Besides, it would not be correct to calculate $\varepsilon'(\omega)$ from equation (6.28) unless the conductivity γ_s were negligible. It has to be calculated for $J_D(t) + J_i(t)$ alone, since only these currents build up a polarization. Only the integral of $J_D(t)$ gives the static dielectric constant due to molecular polarization.

The relative importance of γ_s and $\dot{\Psi}(t)$ in equation (6.28) can be estimated in terms of the order of magnitude of the direct conductivity. A direct conductivity of the order $\gamma_s = 10^{-13}\,\Omega^{-1}\,\text{cm}^{-1}$ makes after one minute of discharge a contribution of the order unity to $\varepsilon''(\omega)$. This contribution is larger than that normally expected from dielectric relaxation, and hence the discharge current is dominated by γ_s. On the other hand, for $\gamma_s < 10^{-15}\,\Omega^{-1}\,\text{cm}^{-1}$ its contribution becomes important only after a much longer time. For total one minute conductivities of less than $10^{-15}\,\Omega^{-1}\,\text{cm}^{-1}$ it becomes increasingly difficult to distinguish between direct conductivity and dielectric relaxation.

Williams[4] describes an experimental arrangement for the measurement of dielectric loss by the measurement of a decay function. The simplest circuit which may be used measures the current by means of a voltage drop across a relatively small resistor in series with the specimen. Williams uses a more sophisticated circuit shown in Fig. 2(a), where the potential across the sample is always kept constant by means of a feedback arrangement which keeps the voltage across R_2 equal and opposite to that across R_1 which is being metered, so that point A is virtually earthed. Furthermore, the design achieves a reduction of the time constant compared with the simple circuit by mounting R_1 in an earthed box as shown in Fig. 2(b), so that the capacitance between A and B is reduced. The switch S_2 is used to protect the amplifier against switching surges. The time constant of the equipment is 0·6 sec as against 18 sec for the commercial d.c. amplifier used.

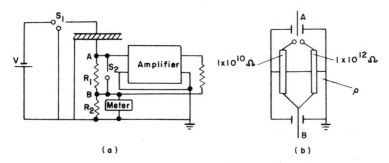

FIG. 2. (a) Circuit for the measurement of decay functions. S_1 is a switch, S_2 a manual make-and-break device. (b) A box for mounting resistors R_1 so that the capacitance between points A and B is kept small. Two resistors are shown; the intercepting plate P is earthed. (From Williams[4])

Williams[4] finds a good fit between the extrapolated step-function data for a polymer and $\varepsilon''(\omega)$ obtained from a.c. data. Reddish[5] maps $\varepsilon''(\omega)$ for several polymers over a wide range of temperatures, and finds very good fit between the step-function and alternating current data. This indicates that equation (6.28) is very useful for normal dielectrics. Reddish[6] recently developed refinements of the mathematical treatment summarized in the present section.

B. ALTERNATING CURRENT MEASUREMENTS FOR FREQUENCIES BELOW 10^7 C/S

The vast majority of dielectric measurements employ alternating current methods. General surveys of these methods, in a wide context, may be found in text books[7]. Critical resumés of techniques are given in publications[8,9] concerned with standard methods. These methods are usually subdivided according to the frequency range used.

The frequency range between 10^{-2} c/s and 10^7 c/s is the most convenient frequency range, where bridge methods may be used, in that the impedance of the capacitor containing the dielectric is balanced against a known combination of discrete resistances, capacitances, or inductances. Many different bridge circuits are covered in the classic book by Hague[10] up to the 1950s. Modern commercial bridges are usually variants of two types of basic circuit, namely the Schering bridge and the transformer ratio bridge. The latter design gained prominence in the decade between 1950 and 1960, for reasons to be discussed later, but the Schering bridge still has the advantage for certain applications. In particular, a variant of the Schering bridge designed by Scheiber[11] can be used for frequencies down to about 10^{-2} c/s, so that there is no frequency gap between direct and alternating current methods.

The most primitive form of Schering bridge is shown in Fig. 3. The simplicity of the circuit facilitates the discussion of experimental complications.

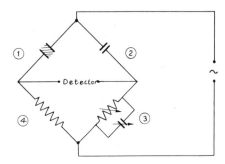

FIG. 3. Schematic circuit of the simplest type of Schering bridge.

The Schering bridge shown is a special case of a bridge with four arms analogous to the Wheatstone bridge originally used for direct current measurements. Such a bridge is balanced if an alternating voltage applied to opposite corners of the rectangle leads to zero voltage across the other pair of corners. When this condition is achieved an equation

$$Z_1^*(\omega)Z_4^*(\omega) = Z_2^*(\omega)Z_3^*(\omega) \tag{6.31}$$

holds between the complex impedances of the four arms. When $Z_1^*(\omega)$ is the impedance of the capacitor which contains the dielectric under investigation it is, according to Chapters 1 and 2 and Appendix I, proportional to $\varepsilon^*(\omega)$ of the dielectric as long as contacts are ideal. Equating of the real and imaginary parts in equation (6.31) yields the complex dielectric constant of the dielectric as a function of Z_2^* to Z_4^* which are the impedances of components or combinations of components which may be known with high accuracy. On the basis of the simple circuit (6.3) one might suppose that bridge measurements could easily be carried out so as to give very high precision because balancing is a null method and alternating currents may be highly amplified. In practice, the circuit shown, assembled casually on a laboratory bench from precision components, will give errors of tan δ of the order 10^{-2}. Higher accuracy demands a skill in bridge design which mounts steeply with the accuracy desired.

6. DIELECTRIC MEASUREMENTS AND THEIR INTERPRETATION

The main reason for the difficulties of bridge design lies in the presence of "stray" capacitances, inductances and capacitances between the different components and between the components and earth. The results obtained with a primitive network such as sketched in Fig. 3 vary quite appreciably with the layout of the components and the wiring. Figure 4 shows a set of strays which produces a particularly puzzling result. If this bridge is balanced and equations (6.31) assumed, neglecting the presence of strays, the resulting tan δ appears negative in apparent defiance of the first law of thermodynamics.

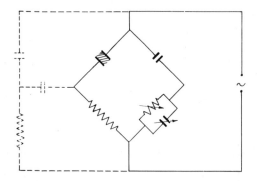

FIG. 4. Simple Schering bridge with stray components (drawn with broken lines) which simulate a negative loss angle.

In other words, the bridge will not balance unless a capacitor C_4 is put in parallel with R_4, even though C_2 is a loss free capacitor. A circuit of this kind may arise if a specimen is placed on an imperfectly insulating support.

A Schering bridge for accurate work minimizes strays by shielding and a guard circuit. That is, all stray capacitances and resistances go as far as possible to an earthed screen and are balanced out by an auxiliary circuit which enables the detector points to be brought close to earth potential. A well designed Schering bridge carefully operated may be used to measure small loss angles with an accuracy of $\pm 10^{-6}$ in increments of tan δ.

The Schering bridge is at its best at frequencies below 10–100 kc/s but can be used up to frequencies of, say, 500 kc/s. It is particularly suitable for work at high voltages, but is not suitable for very large losses, say tan $\delta > 1$, although corrections may be calculated.

The basic feature of a transformer ratio bridge is most effectively shown in an early form of the design which has been further simplified in Fig. 5. This bridge is analogous to a Wheatstone bridge, but two arms consist of the two windings of a transformer. When a voltage is applied to the bridge the magnetic flux is common to both windings. This has the consequence that any stray admittance across the winding on the specimen side is "reflected" by the transformer to appear present also on the standard side. In first approximation the strays cancel against their own reflection. Hence a transformer bridge is less affected by strays to earth than a Schering bridge. Later designs

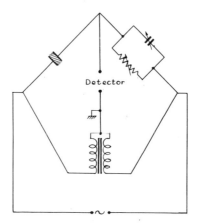

FIG. 5. Schematic circuit of the simplest type of transformer bridge.

of this type of bridge use two transformers, thus isolating the bridge from the source and detector, as shown in Fig. 6. Tappings on one or both transformers may be used to balance a wide range of values of the unknown against standards of moderate range. Strays to earth may be minimized by connecting shields of cables etc. to the neutral.

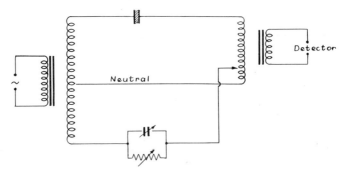

FIG. 6. Schematic circuit of a commercial transformer ratio bridge.

The accuracy of transformer ratio bridges may be as high as for Schering bridges, but the highest precision is likely to be confined to a less wide frequency range than for a Schering bridge. The upper frequency limit of the use of transformer ratio bridges is set by the inductance of the leads to the specimen. A limit of a few Mc/s is typical for the use of a commercial bridge with short leads, a capacitance $< 10^{-10}$ F and a permitted error of less than 2% in increments of tan δ. Details of design are important here.

The performance of a bridge depends on the oscillators and detectors with which it is used. Overtones of the signal frequency produced by defects in the oscillator or non-ohmic contacts elsewhere give a background of noise in the balance condition. A sophisticated measuring equipment usually uses filters in its detector to isolate the frequency used. By this procedure the bridge assembly is apt to frustrate the observation of non-linear phenomena in the specimen.

Even with a well designed bridge it is distinctly exacting to measure power factors reproducibly to $\pm 3 \times 10^{-5}$. For accurate work it is best not to trust the design of any bridge too much but to calibrate the apparatus by replacing the specimen by a "dummy" combination of a good loss-free variable capacitor and various resistors. By this means systematic errors may be detected and allowed for.

C. ALTERNATING CURRENT MEASUREMENTS FOR FREQUENCIES ABOVE 1 MC/S

For frequencies higher than a few Mc/s representation of a circuit by discrete circuit elements and by wires free of inductance and capacitance becomes gradually unrealistic. For higher frequencies circuits are considered as distributed impedances, and the specimen incorporated accordingly.

For the frequency range up to 10^9 c/s resonant circuits may be used, the specimen being combined with a known inductance. Resonant circuits are used down to low frequencies (about 1 kc/s) in the form of Q meters. Such a meter uses essentially a series combination of C, L and R. The energy absorption exhibits a resonant peak and the width of this peak $2\Delta\omega$ is measured. $\Delta\omega$ is the difference between the frequency where the energy loss is a maximum and the frequency (on either side of the peak) where it is reduced to half the maximum. For a simple series circuit the Q value is given by

$$Q = \frac{\omega_0}{2\Delta\omega} = \frac{\omega_0 L}{R} = \frac{1}{\omega_0 CR}. \tag{6.32}$$

In so far as the capacitor is the only lossy component Q is thus inversely proportional to tan δ of the dielectric at ω_0. Q meters which use leads from instrument to specimen may be inexpensive and versatile, but they are generally not suitable for precision measurements of loss angle.

A resonance circuit of high accuracy has been developed by Hartshorn and Ward[8]. Here the specimen is inserted into a "head" the capacitance of which is varied by micrometer screws, so that no leads to the specimen are involved. This method allows the measurement of loss angles down to 2×10^{-5} at

10–20 Mc/s. It operates between 10 kc/s and 100 Mc/s. A variant of this method works up to 1000 Mc/s. The Hartshorn-Ward method needs separate heads for solids and liquid specimens. It is very difficult to adapt this method for use over more than a narrow range of temperatures.

A frequency of 300 Mc/s corresponds to a wavelength of a metre, and for frequencies above this range the shape and size of the measuring assembly is important in relation to the wavelength. Because of this, different specimens have to be used for not so very different frequencies. Measurements in this "microwave" region are discussed in references 8 and 9 and in text books[12]. Cavity resonators in the 9000–36,000 Mc/s range are able to measure tan δ values between 2×10^{-5} and 3×10^{-2}, and accuracies to better than 10^{-4} can be reached over the whole range, by a suitable choice of equipment.

D. ELECTRODE ASSEMBLIES

In the discussion so far it was assumed that the material under test fills a capacitor so that it makes perfect contact with perfectly clean electrodes, that it fills the capacitor without leaving voids and that there is no fringing of the electrical field at the edges of the capacitor. In practice[8], the last two of these conditions are of importance mainly with regard to accurate permittivity measurements. The first two are important also for measurements of loss angle.

For frequencies from d.c. up to 1 or 10 Mc/s all measurements on a material may be carried out in the same electrode assembly, which ensures that most sources of error are the same throughout. For solid specimens errors in loss angle are most likely at the high frequency end of this region, due to bad contact. At a frequency of 1 Mc/s a resistance of 1 Ω in series with 10^{-9} F introduces an error of $6\cdot3 \times 10^{-3}$ in loss angle. Hence a poor electrode such as badly applied colloidal carbon may introduce quite large errors. For a very bad electrode this error is easily recognized since the apparent permittivity at the highest frequencies drops towards unity, a value impossibly low for a solid insulating material. Otherwise the measured tan δ is easily mistaken for the foothills of a relaxation peak at high frequencies.

Measurements of loss on liquid insulating materials at low frequencies are made difficult by the possibility of contamination by dirty surfaces. Test cells suitable for measurements on liquids are described[8] in document L/S9 together with electrodes for solid surfaces.

The different methods for frequencies beyond 10^8 Mc/s or so demand specific shapes for the specimen, so that different errors are introduced at different frequencies. In some methods the fit of electrodes is more important than others. Altogether, the measurement of tan δ over the whole a.c. frequency range is a laborious process, and this explains why relatively few such measurements have been done. Most investigators prefer to use frequencies and methods where leads can be used and to alter the temperature, relying on a theoretically plausible relationship between temperature and relaxation time (see equation 4.47) to approach the shortest times by a reduction in temperature. The use of electrically shielded ovens presents little difficulty, but some care is needed at low temperatures to avoid the formation of conducting paths in unsuitable places due to condensed moisture. Besides, the cooling of a capacitor inside an evacuated, or even a merely desiccated, enclosure is made relatively slow by the fact that bad conductors of electricity are generally also bad conductors of heat.

REFERENCES

1. J. Hopkinson (1901). Original Papers, Vol. 2, p. 119. Cambridge University Press, London.
2. K. S. Cole and R. H. Cole (1942). *J. Chem. Phys.*, **10**, 98.
3. B. V. Hamon (1952). *Proc. Instn. elect. Engrs.*, **99**, Part IV, Monograph 27.
4. G. Williams (1963). *Polymer*, **4**, 27.
5. W. Reddish (1962). "Pure and Applied Chemistry." Vol. 5, p. 723. Butterworth, London. Also (1958). *Soc. chem. Ind. Symp.*,
6. W. Reddish. Personal Communication.
7. A. R. von Hippel (ed.) (1954). "Dielectric Materials and Applications." Chapter 2. Chapman and Hall, London.
8. Electrical Research Association (1963). Technical Report L/S9. "Methods of Testing Permittivity and Loss Tangent of Dielectric Materials."
9. American Society for Testing and Materials (1966). Book of ASTM Standards, Part 29, "Electrical Insulating Materials."
10. B. Hague (1957). "Alternating Current Bridge Methods", 5th ed. Pitman and Sons, London.
11. D. J. Scheiber (1961). *J. Res. natn. Bur. Stand.*, 65C, 23–32.
12. H. M. Barlow and A. L. Cullen (1950). "Microwave Measurements." Constable, London.

CHAPTER 7

Empirical Methods for the Evaluation of Dielectric Measurements

Dielectric measurements provide data on permittivity and dielectric loss as a function of frequency and temperature. These data are conveniently displayed in the form of graphs for the real and imaginary parts of the permittivity as a function of $\log_{10} \omega$. In principle such graphs could be used for the numerical computation of distribution functions $y(\tau)$ but in practice the information is hardly ever complete and precise enough to warrant this labour. The data are generally evaluated in terms of certain empirical relationships—with more or less firm foundations in theory—which allow their characterization by means of a small number of parameters.

A. METHODS OF DISPLAYING RELAXATION PEAKS

The evaluation of experimental data is much facilitated by certain graphical methods of display, which permit the derivation of parameters by geometrical construction. The earliest and most used of these methods consists of plotting $\varepsilon''(\omega)$ for a certain frequency against $\varepsilon'(\omega)$ at the same frequency, in cartesian coordinates or in the complex plane. This diagram may be called the complex locus diagram or Argand diagram and was applied to dielectrics by Cole and Cole[1]. It is often called the Cole–Cole plot or arc plot.

For a dielectric with a single relaxation time the Cole–Cole plot is a semi-circle. A simple evaluation of the Debye equations (2.21) and (2.22) shows that the equation between the real and imaginary part of the dielectric constant is the equation of a circle

$$\left(\varepsilon'(\omega) - \frac{\varepsilon_s + \varepsilon_\infty}{2}\right)^2 + (\varepsilon''(\omega))^2 = \tfrac{1}{4}(\varepsilon_s - \varepsilon_\infty)^2. \tag{7.1}$$

The Cole–Cole plot provides therefore an elegant method of finding out whether a system has a single relaxation time. This plot is also useful for the characterization of different types of distribution function, and is widely applied.

The arc plot of a dielectric with a single relaxation time is shown in Fig. 1. A given point on the semi-circle corresponds to a given frequency while the summit corresponds to $\omega\tau = 1$ (see also Fig. 2 a and b). The plot has the disadvantage that ω does not appear in it. Any material with a single relaxation time, characterized by ε_s and ε_∞ gives the same arc plot. In reporting results it is therefore essential to supply the magnitude of τ in addition to the arc plot. The same consideration applies to all representations which plot $\varepsilon''(\omega)$ as a function of $\varepsilon'(\omega)$, but the frequency reference is not always as simple as in the Debye case.

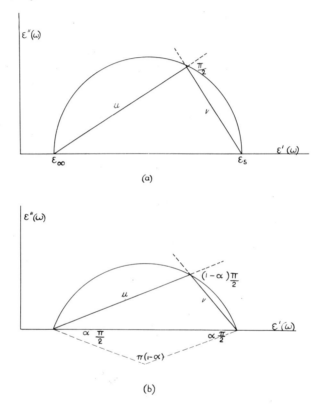

FIG. 1. (a) Arc plot for a Debye dielectric; (b) Arc plot for a dielectric with a Cole–Cole distribution characterized by the parameter α (from Cole and Cole[1]).

The plot in Fig. 1(a) can be considered in cartesian coordinates but its treatment is mathematically more elegant in the complex plane. A point on the semi-circle defines two vectors **u** and **v** as shown. By virtue of the construction

$$\mathbf{u} - \mathbf{v} = \varepsilon_s - \varepsilon_\infty \tag{7.2}$$

while

$$\mathbf{u} = \varepsilon'(\omega) - \varepsilon_\infty + i\varepsilon''(\omega) \tag{7.3}$$

7. EMPIRICAL METHODS FOR EVALUATION OF DIELECTRIC MEASUREMENTS

by definition. If the Debye equations apply

$$\mathbf{v} = \mathbf{u}i\omega\tau \tag{7.4}$$

which signifies that the two vectors are at right angles to each other, since multiplication by i signifies rotation by $\pi/2$ in the complex plane.

Cole and Cole generalize the representation of a Debye dielectric by a circular arc plot in the complex plane so that it applies to a certain type of distributions of relaxation times. They retain the vectors $\mathbf{u}$ and $\mathbf{v}$ and equations (7.2) and (7.3) but demand that $\varepsilon^*(\omega)$ should be so defined that

$$\mathbf{v} = \mathbf{u}(i\omega\tau_0)^{1-\alpha} \tag{7.5}$$

where τ_0 and α are constants. The geometrical representation of this assumption is shown in Fig. 1(b). For any point on the arc defined by plotting $\varepsilon''(\omega)$ as a function of $\varepsilon'(\omega)$ the angle between $\mathbf{u}$ and $\mathbf{v}$ is $(1-\alpha)\pi/2$. This is the case for two secants of a circle where the radii drawn from the centre of the circle subtend an angle $\pi(1-\alpha)$ as shown, and the construction of the circle follows from the diagram. The corresponding complex dielectric constant is given by

$$\varepsilon^*(\omega) = (\varepsilon_s - \varepsilon_\infty)\frac{1}{1+(i\omega\tau_0)^{1-\alpha}}. \tag{7.6a}$$

The distribution of relaxation times which correspond to this $\varepsilon^*(\omega)$ will be discussed in the next section. For $\alpha = 0$ equation (7.6) reduces to the Debye equations.

Equation (7.6a) may be separated into its real and imaginary part, using the identity

$$i^\alpha \equiv e^{i\frac{\pi}{2}\alpha}. \tag{7.7}$$

This gives

$$\varepsilon'(\omega) - \varepsilon_\infty = \tfrac{1}{2}(\varepsilon_s - \varepsilon_\infty)\left(1 - \frac{\sinh(1-\alpha)s}{\cosh(1-\alpha)s + \cos\tfrac{1}{2}\alpha\pi}\right) \tag{7.6b}$$

$$\varepsilon''(\omega) = \frac{\tfrac{1}{2}(\varepsilon_s - \varepsilon_\infty)\cos\tfrac{1}{2}\alpha\pi}{\cosh(1-\alpha)s + \sin\tfrac{1}{2}\alpha\pi} \tag{7.6c}$$

where $s = \log_e \omega\tau_0$.

The representation of dielectric data in terms of $\varepsilon^*(\omega)$ is convenient where the steady conductivity does not contribute significantly to $\varepsilon''(\omega)$. Where this conductivity is appreciable Grant[2] uses a plot in terms of a complex conductivity defined so that

$$\sigma'(\omega) + i\sigma''(\omega) = i\omega(\varepsilon'(\omega) - i\varepsilon''(\omega)). \tag{7.8}$$

Figure 2, reproduced from Grant, shows a comparison of conductivity and permittivity plots for four equivalent circuits. The complex conductivity $\sigma^*(\omega)$ is here defined so that it is 4π times the conductivity in e.s.u., and the steady value is related to that quoted in equation (6.24) so that $\varepsilon_0\sigma_0 = \gamma_s$ (see Appendices I and II). The figures show that a steady conductivity distorts the permittivity plot while the presence of ε_∞ distorts the conductivity plot. The

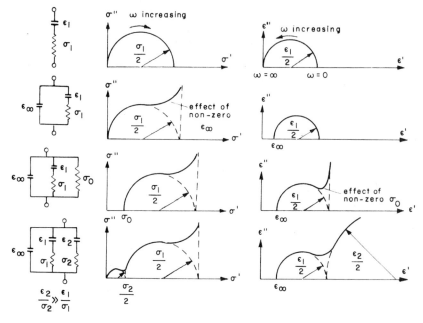

FIG. 2. Diagrams of complex conductivity (σ'' against σ') and complex permittivity (ε'' against ε') for four equivalent circuits (from Grant[2]).

latter distortion can be eliminated[2] by using $\varepsilon'(\omega) - \varepsilon_\infty$ to formulate $\sigma''(\omega)$. The conductivity plot gives an unambiguous value σ_0 only if the steady conductivity is fairly large and not significantly voltage dependent, as explained in Section A of Chapter 6.

Scaife[3] suggests another method of display which is based on a theoretical consideration of electrostatic interaction. Scaife argues that a spherical macroscopic specimen *in vacuo* is the most suitable starting point for the analysis of separate mechanisms of relaxation. The complex polarizability of such a specimen may be derived as

$$\alpha^*(\omega) = \frac{\varepsilon^*(\omega) - 1}{\varepsilon^*(\omega) + 2} \tag{7.9}$$

from the measured dielectric data. The theoretical significance of this

7. EMPIRICAL METHODS FOR EVALUATION OF DIELECTRIC MEASUREMENTS

polarizability, as a measure of the fluctuation of the polarization of a spherical specimen, is discussed in Section B of Chapter 8.

The polarizability plot is useful for a reason demonstrated by the last diagram of Fig. 2, for a dielectric with a slow and a fast mechanism of relaxation. For the slow process the effective value of ε_∞ is near to the static value ε_1 of the fast process. As a consequence the static value ε_2 of the slow process is enhanced by the response of the fast process. This enhancement takes place in the case of a measurement of ε in a plate capacitor, but would not take place if the fluctuations of the polarization in a spherical specimen were measured. Scaife derives the data for the sphere by a calculation which involves only macroscopic electrostatics.

For a single relaxation time $\alpha''(\omega)$ as a function of $\alpha'(\omega)$ is a semi-circle. It is a circular arc plot also for the case that $\varepsilon^*(\omega)$ is given by equation (7.6). The polarizability plot has a theoretical background in general, but is in practice most useful for an experimental situation like the one sketched in the last diagram of Fig. 2, where the polarizability plot would give improved resolution[3].

B. DISTRIBUTIONS OF RELAXATION TIMES

A number of distribution functions are in use which describe $y(\tau)$ by means of a limited number of parameters.

Before discussing distribution functions which may be defined for the simple evaluation of dielectric data it will be helpful to discuss a function which is physically simple. Such a function is treated in detail by Fröhlich[4]. He uses a dynamic model where τ depends on T according to equation (4.47) and assumes that the activation energy varies between the constant values W_0 and $W_0 + v$. Thus the relaxation time varies between

$$\tau_0 = A\, e^{\frac{W_0}{kT}} \tag{7.10}$$

$$\tau_1 = A\, e^{\frac{W_0 + v_1}{kT}}. \tag{7.11}$$

Fröhlich assumes that the polarizable units are evenly distributed in terms of activation energy. That is, dipoles in an energy interval dv contribute to the static dielectric constant a term

$$d\varepsilon_s = (\varepsilon_s - \varepsilon_\infty)\frac{dv}{v_1} \tag{7.12}$$

independently of the position of the interval dv. This assumption defines $y(\tau)$ since according to equations (5.9) and (5.10)

$$d\varepsilon_s = y(\tau)\, d\tau$$

$$= y(\tau)\frac{d\tau}{dv}\, d\tau$$

and differentiation of equation (7.11) (with variable v) gives

$$d\varepsilon_s = y(\tau)\frac{\tau}{kT}dv. \tag{7.13}$$

Comparison of equations (7.12) and (7.13) leads to

$$y(\tau) = (\varepsilon_s - \varepsilon_\infty)\frac{kT}{v_1}\frac{1}{\tau} \quad \text{for } \tau_0 < \tau < \tau_1 \tag{7.14}$$

$$y(\tau) = 0 \quad \text{for } \tau < \tau_0 \text{ and } \tau > \tau_1$$

for the distribution functions defined according to equations (5.10) and (5.11). In terms of the notation of Appendix V this distribution function can be expressed even more simply as

$$F(s) = \frac{kT}{v_1} \quad \text{for } 0 < s < \frac{v_1}{kT} \tag{7.15}$$

where $s = \log\dfrac{\tau}{\tau_0}$. The real and imaginary parts of $\varepsilon^*(\omega)$ which correspond to this distribution are given by[4]

$$\varepsilon'(\omega) - \varepsilon_\infty = (\varepsilon_s - \varepsilon_\infty)\left(1 - \frac{kT}{2v_1}\log\frac{1 + \omega^2\tau_0^2\, e^{\frac{2v_1}{kT}}}{1 + \omega^2\tau_0^2}\right) \tag{7.16a}$$

$$\varepsilon''(\omega) = (\varepsilon_s - \varepsilon_\infty)\frac{kT}{v_1}(\tan^{-1}\omega\tau_0\, e^{\frac{v_1}{kT}} - \tan^{-1}\omega\tau_0) \tag{7.16b}$$

fairly complicated formulae which reduce to the Debye equations for $v_1 = 0$. The function $\varepsilon''(\omega)$ is a maximum for

$$\omega_m = \frac{1}{\tau_0}e^{-\frac{1}{2}\frac{v_1}{kT}} = \frac{1}{\sqrt{(\tau_0\tau_1)}}. \tag{7.17}$$

Figure 3 shows the dielectric loss calculated from equation (7.16b) for different widths of the distribution, characterized by $\sqrt{(\tau_1/\tau_0)}$. The graph is calculated for $\varepsilon''(\omega)/\varepsilon''(\omega)_m$ where ω_m is given by equation (7.17). The formula for calculating Fig. 7.3 is given by Fröhlich[4] as

$$\frac{\varepsilon''(\omega)}{\varepsilon''(\omega_m)} = \frac{\tan^{-1}\dfrac{\omega}{\omega_m}\sqrt{\left(\dfrac{\tau_1}{\tau_0}\right)} - \tan^{-1}\dfrac{\omega}{\omega_m}\sqrt{\left(\dfrac{\tau_0}{\tau_1}\right)}}{\tan^{-1}\sqrt{\left(\dfrac{\tau_1}{\tau_0}\right)} - \tan^{-1}\sqrt{\left(\dfrac{\tau_0}{\tau_1}\right)}}. \tag{7.16c}$$

7. EMPIRICAL METHODS FOR EVALUATION OF DIELECTRIC MEASUREMENTS

Figure 3 shows that the loss peaks become wider and wider as v_1 increases. In the ultimate limit the distribution would become

$$y(\tau) = \frac{\text{const.}}{\tau} \tag{7.18}$$

with τ varying from zero to infinity. Introduction of this function into equation (5.14) leads to a constant $\varepsilon''(\omega)$ over the whole frequency range, in view of equation (5.15). Such a frequency dependence of the dielectric loss does occur in practice over a very wide frequency range and will be discussed in Chapter 13.

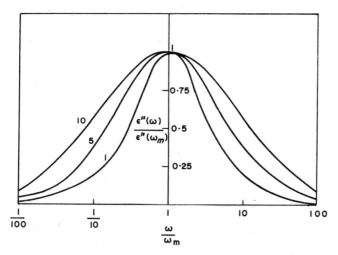

FIG. 3. $\varepsilon''(\omega)$ according to equation (7.16c) for three values of the parameter $\sqrt{\dfrac{\tau_1}{\tau_0}}$ (from Fröhlich[4]).

Fröhlich's distribution function (7.14) is essentially a teaching model. There is no reason why activation energies should be evenly distributed over a small range of values while higher or lower values of W should be absent. However, Fröhlich's distribution can in principle be easily generalized by assuming that most activation energies have a median value while deviant values are increasingly unlikely. The theoretically simplest assumption of this kind is a Gaussian distribution of activation energies. This means in terms of the variable s used in equation (7.15)

$$F(s) = \frac{b}{\sqrt{\pi}} e^{-b^2 s^2} \tag{7.19}$$

where s now varies from zero to infinity.

Unfortunately the Gaussian distribution does not lend itself to simple evaluation. However, the function $\varepsilon^*(\omega)$ in equation (7.6) has been designed by Cole and Cole for simple evaluation. The distribution function which corresponds to it (see equations 5.13, 5.14 and Appendix V) is given by[1]

$$F(s) = \frac{1}{2\pi} \frac{\sin \alpha \pi}{\cosh (1 - \alpha)s - \cos \alpha \pi}. \qquad (7.20)$$

This function is in practice not so very different from a Gaussian distribution. Figure 4 shows[1] that equations (7.19) and (7.20) can be made to coincide fairly closely by a suitable choice of the parameters b and α. The Cole–Cole distribution falls off more slowly towards extreme values of s than the Gaussian distribution.

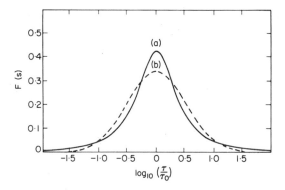

FIG. 4. Distributions of relaxation times $F(s)$ where $s = \log_e (\tau/\tau_0)$. Calculated (a) for a Cole–Cole distribution (equation 7.20) with $\alpha = 0.23$, and (b) for a Gaussian distribution (equation 7.19) with $b = 0.6$ (from Cole and Cole[1]).

The Cole–Cole distribution is a useful representation for many experimental results, and this seems reasonable in view of its similarity to a Gaussian distribution. For small $1 - \alpha$ it is also well compatible with the empirical equation (6.14) for measured decay functions. Before discussing this relationship it will be useful to note that for a decay function given by equation (6.14), with m within the limits defined, the loss is approximately constant. In view of equations (6.28) and (6.30)

$$\varepsilon''(\omega) = \frac{1}{\varepsilon_0} \dot{\Psi}\left(\frac{0.63}{\omega}\right)$$

$$= \frac{C}{\varepsilon_0 0.63}. \qquad (7.21)$$

However, it has been shown that equation (7.18) implies a constant $\varepsilon''(\omega)$. This means that $y(\tau)$ in equation (7.18) represents $\psi(t)$ in equation (6.14) for $0.3 < m < 1.2$ in fair approximation.

Cole and Cole[5] derive $\Psi(t)$ from equation (7.6). Williams[6] examines the result of this evaluation with regard to the validity of equation (6.28), for the case that $\varepsilon^*(\omega)$ is given by equation (7.6) over the whole frequency range. He finds that for a wide distribution with about $1 < \alpha < 0.7$ the Cole–Cole equation (7.6) and equation (6.14) can be fitted to each other whether the time in $\Psi(t_1)$ is larger or smaller than the time parameter τ_0 in equation (7.6). For a narrow distribution the fit is poor for $t_1/\tau_0 > 0$ which is the case of interest for very low frequencies. In other words, a Cole–Cole function with a constant α fits over the whole frequency range only in the case of a very wide distribution. The fit, in this respect, is probably better for the Cole–Cole distribution than for a Gaussian distribution, in view of Fig. 4. This implies that equation (7.6) seems a fortunate empirical compromise as a representation of a distribution function using only two parameters.

The Cole–Cole distribution function is not the only one which combines convenient graphical representation with a good empirical fit by means of a small number of parameters. Fuoss and Kirkwood[7], in a classical early paper, derive the equation

$$\varepsilon''(\omega) = \frac{\varepsilon_m''}{\cosh as} \qquad (7.22)$$

where a is a constant, $s = \log \omega/\omega_m$ and ω_m and ε_m'' refer to the centre of the (symmetrical) distribution. This expression gives a straight line of slope a if $\cosh^{-1} \varepsilon_m''/\varepsilon''(\omega)$ is plotted against $\log \omega$, a relationship which is well obeyed by the polyvinylchloride compositions measured by the authors. The plot in question has the advantage that it is convenient for the determination of the central frequency ω_m. Cole[8] compares expression (7.22) with expression (7.6), on the straight line plot, and finds that they give results which are distinguishable only if a wide frequency range is used. The Fuoss–Kirkwood function corresponds to a logarithmic distribution function

$$F(s) = \frac{a}{\pi} \frac{\cos\left(\frac{a\pi}{2}\right) \cosh(as)}{\cos^2\left(\frac{a\pi}{2}\right) + \sinh^2(as)}. \qquad (7.23)$$

The distribution functions discussed so far are symmetrical as functions of log τ/τ_0 and represent a dependence $\varepsilon''(\omega)$ which is similar to a widened Debye curve. However, a number of materials give unsymmetrical loss curves. These can be represented conveniently by an expression due to Davidson and Cole[9]

$$\varepsilon^*(\omega) - \varepsilon_\infty = \frac{\varepsilon_s - \varepsilon_\infty}{(1 + i\omega\tau_0)^\beta} \qquad (7.24)$$

where τ_0 and β are constants. The function gives a characteristic shape (see Fig. 5) both on the permittivity and the polarizability plot which is easily recognized as a "skewed arc". The distribution function for this case is

$$F\left(\frac{\tau}{\tau_0}\right) = \frac{\sin \beta\pi}{\pi} \left(\frac{\tau}{\tau_0 - \tau}\right)^\beta \quad \text{for} \quad \tau < \tau_0 \qquad (7.25)$$

$$= 0 \qquad \text{for} \quad \tau > \tau_0.$$

The skewed arc looks similar to a semi-circle at its low frequency end while at the high frequency end it resembles a circular arc plot. Figure 5 shows how a

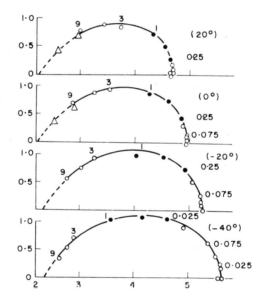

Fig. 5. Skewed arc plots for *n*-octyl iodide. Frequencies are in units of 10^9 c/s (from Cole and Mopsik[10]).

complete representation of $\varepsilon''(\omega)$ data on an arc plot can be achieved by writing the frequencies of measurement next to the measured points, and by giving data as a function of temperature.

Davidson and Cole[9] describe graphical methods for the derivation of the parameters β and τ_0 and also give the decay function which corresponds to equations (7.24) and (7.25).

Many distribution functions other than those quoted here have been defined. An appendix to the monograph by Gross (Reference 2 of Chapter 5) lists a number of distribution functions together with their bibliographic references.

C. THE INTERPRETATION OF WIDE LOSS PEAKS IN TERMS OF ACTIVATION ENERGIES

A dielectric relaxation time τ refers in the last resort to the regression time of fluctuations of the polarization, and is accordingly very dependent on temperature (see Chapter 4). The relationship for a single τ is given by equation (4.47) or (4.48). When we consider a distribution of relaxation times either the pre-exponential factor or the activation energy, or both, may be distributed. Nevertheless, we wish to draw conclusions regarding activation energies and pre-exponential factors from the shift with temperature of wide relaxation peaks.

Macdonald[11] points out that a distribution in W alone which is invariant with temperature is not compatible with several of the distribution functions which have been designed for convenience in evaluation and display. Fröhlich's model is temperature invariant in this sense. Equation (7.15) is a distribution function in terms of s which is normalized so that its integral from minus infinity to plus infinity equals unity (see equation V.8). If one defines a distribution function in terms of W which is normalized so that

$$\int_0^\infty K(W)\,dW = 1 \tag{7.26}$$

then

$$K(W) = \frac{1}{kT} F(s) \tag{7.27}$$

and $K(W)$ for Fröhlich's model is constant, in view of equation (7.15). This implies that τ_1/τ_0 decreases with rising temperature as may be seen from equations (7.10) and (7.11), and that the width of the loss peaks decreases, as may be seen from Fig. 3.

A Gaussian distribution of activation energies is closely related to Fröhlich's distribution and also temperature invariant in that

$$K(W) = \frac{1}{\sigma\sqrt{2\pi}} \exp\left(-\frac{1}{2}\left(\frac{W - W_0}{\sigma}\right)^2\right). \tag{7.28}$$

The parameter σ which characterizes the width of this temperature invariant distribution may be compared with the parameter b in equation (7.19) which characterizes the width of the distribution in terms of log τ. It may be seen that

$$b = \frac{\sigma\sqrt{2}}{kT} \qquad (7.29)$$

which signifies that the distribution in terms of relaxation times narrows with rising temperature.

A narrowing of loss peaks with rising temperature is often observed. It seems plausible to expect that distribution functions should behave with regard to temperature as in the case of the two models described above because activation energies are an attribute of a solid of liquid structure. One would expect their distribution to be little dependent on temperature, as a rule.

The form of the Cole–Cole distribution function (7.20) is such that it cannot be described by a temperature independent function $K(W)$. The function $K(W)$ can be made approximately temperature independent for a very wide or a very narrow distribution. However, for distributions of intermediate width it would seem that the Cole–Cole plot should not be able to retain a circular arc shape over a wide range of temperatures, unless the distribution of activation energies changed with temperature. This is an odd situation, since the Cole–Cole function has been found generally useful for the description of experimental data.

Macdonald considers also the Fuoss–Kirkwood distribution and the skewed arc, and does not find them much more satisfactory with regard to a temperature independent distribution of activation energies. The fact that the discrepancy in question has so far not troubled experimentalists is not as surprising as it may seem. Precise (or even approximate) data for $\varepsilon^*(\omega, T)$ are rarely available over a wide range of frequencies and temperatures, for materials where no changes of structure with temperature intervene. In practice, most data are barely sufficient to determine one parameter beyond a relaxation time reference τ_0.

While the temperature dependence of the distribution of activation energies is a somewhat tricky problem, the evaluation of average activation energies from $\varepsilon''(\omega, T)$ still remains worth while. As a rule, such activation energies have been determined from the shift of the maximum of $\varepsilon''(\omega)$ with temperature. However, this method is not always advantageous since measurements at frequencies over 10 mc/s are often inconvenient, the peak position of a wide curve may not be too well defined, and the peak may not be symmetrically placed.

Read and Williams[12] derive activation energies from curves of ε'' measured as a function of temperature at a given frequency ω. This procedure gives a complete loss curve if at the highest temperature T_1 chosen $\varepsilon(\omega) = \varepsilon_s(T_1)$,

7. EMPIRICAL METHODS FOR EVALUATION OF DIELECTRIC MEASUREMENTS

while at the lowest temperature $\varepsilon(\omega) = \varepsilon_\infty(T_2)$. The reciprocal temperature, $1/T$ has to be used as variable rather than T itself.

For a Debye dielectric where the dependence of ε_s on T may be neglected and equation (4.47) holds with A constant we have exactly

$$(\varepsilon_s - \varepsilon_\infty)\frac{\pi R}{2W} = \int_0^\infty \varepsilon''\left(\frac{1}{T}\right) d\left(\frac{1}{T}\right) \tag{7.30}$$

where R is the gas constant and W is measured in calories/mole. Williams and Read show that the same equation applies approximately for the Fuoss–Kirkwood, Cole–Cole and Cole–Davidson distribution, if we assume the same activation energy W for all relaxation times of a distribution. This assumption means, of course, that the distribution is due to a variation of the pre-exponential factor A.

In general ε_s depends on T, and neglecting dipolar interaction we may put

$$\varepsilon_s(T) - \varepsilon_\infty = \frac{\text{const.}}{T}. \tag{7.31}$$

With the assumption of a single W, equation (7.31) still leads to equation (7.30), with the proviso that $\varepsilon_s(T_m)$ is measured, i.e. the static value for that temperature for which ε'' has a maximum as a function of T, and for constant ω.

When the postulate of a single activation energy W is dropped, equation (7.30) may still be used, but now it measures the average value of $1/W$.

As an alternative to equation (7.30) an average W may be derived from a half width of the curve $\varepsilon''(1/T)$. For a single relaxation time τ the Debye equation may be written in the form

$$\varepsilon''(\omega\tau) = 2\varepsilon''_m \frac{\omega\tau}{1 + \omega^2\tau^2} \tag{7.32}$$

where ε''_m is the maximum loss factor. This curve has two values $\omega\tau$ for which

$$\varepsilon''(\omega\tau) = \frac{1}{r}\varepsilon''_m \tag{7.33}$$

and for these

$$(\omega\tau)_{1,2} = r \pm \sqrt{(r^2 - 1)}. \tag{7.34}$$

In generalizing this expression for use with the $\varepsilon''(1/T)$ curve, we find that a relationship like (7.33) holds for two values of $1/T$, symmetrically placed about the peak value $1/T_m$. The width of the curve

$$\Delta r = \frac{1}{T_r} - \frac{1}{T_m} \qquad (7.35)$$

is a function of W which for a Debye curve is

$$W = \frac{R}{\Delta r} \log_e \frac{r + \sqrt{(r^2 - 1)}}{r - \sqrt{(r^2 - 1)}} \qquad (7.36)$$

where R is the gas constant.

Expression (7.34) is modified according to the form of the distribution of relaxation times. For the Cole–Cole distribution r has to be made equal 2 and

$$W = \frac{R}{\Delta_2} \log_e \left(\frac{x + \sqrt{(x^2 - 1)}}{x - \sqrt{(x^2 - 1)}} \right) \qquad (7.37)$$

where

$$x = 2 + \sin\left(\alpha \frac{\pi}{2}\right)$$

and α is the parameter in equation (7.20).

For the Fuoss–Kirkwood distribution

$$W = \frac{R}{a\Delta r} \log_e \frac{r + \sqrt{(r^2 - 1)}}{r - \sqrt{(r^2 - 1)}} \qquad (7.38)$$

where a is the parameter in equation (7.22).

The formulae given assume that the distribution of relaxation times is temperature independent. Read and Williams briefly discuss the case when this is not so.

Since equation (7.30) holds approximately for all distributions while (7.37) and (7.38) and analogous expressions hold for specific distributions, comparison of the two methods may be used as a check on the type of distribution present. Read and Williams give data on dielectrics obeying the Fuoss–Kirkwood distribution which indicate good agreement between the usual (shift) method and the two new methods for the determination of average activation energies.

In conclusion, most of the methods described in this chapter put a somewhat artificial interpretation on a more or less shapeless mass of data. Evidence reported only in terms of the parameters of any of the empirical formulae leaves out of account a varying amount of detail which may at some stage prove relevant. In particular, none of the formulae discussed is really suitable for a detailed investigation of the problem whether distributions of relaxation times change with temperature.

However, the empirical expressions are valuable in classifying dielectrics and in comparing materials belonging to a given group.

REFERENCES

1. K. S. Cole and R. H. Cole (1941). *J. Chem. Phys.*, **9**, 341.
2. F. A. Grant (1958). *J. Appl. Phys.*, **29**, 76.
3. F. K. P. Scaife (1963). E.R.A. report ref. No. L/T416. *Proc. Phys. Soc.*, **81**, 124.
4. H. Fröhlich (1958). "Theory of Dielectrics." Oxford University Press, London.
5. K. S. Cole and R. H. Cole (1942). *J. Chem. Phys.*, **10**, 98.
6. G. Williams (1962). *Trans. Faraday Soc.*, **58**, 1041.
7. R. M. Fuoss and J. G. Kirkwood (1941). *J. Am. Chem. Soc.*, **63**, 385
8. R. H. Cole (1955). *J. Chem. Phys.*, **23**, 493. Also (1961). "Progress in Dielectrics." Vol. 3, p. 49. Heywood, London.
9. D. W. Davidson and R. H. Cole (1951). *J. Chem. Phys.*, **19**, 1484.
10. R. H. Cole and F. J. Mopsik (1966). *J. Chem. Phys.*, **44**, 1015.
11. J. R. Macdonald (1962). *J. Chem. Phys.*, **36**, 345.
12. B. E. Read and G. Williams (1961). *Trans. Faraday Soc.*, **57**, 1979.

CHAPTER 8

Electrostatic Interaction

The previous chapters, with the exception of the final paragraph of Chapter 2, neglect electrostatic interaction. That is, dielectric relaxation is considered subject to the condition that the long range electrostatic interaction between dipoles or other polarizable elements is negligible. The present chapter examines the effect of electrostatic interaction on dielectric relaxation.

Electrostatic interaction is a difficult subject for two reasons. In the first place field theory is quite complicated even for macroscopic bodies which are considered as if the matter within them were continuous and isotropic. Secondly, real matter is made up of units of atomic size which move spontaneously, so that all macroscopic properties of an assembly of such units fluctuate (see Chapter 4). The first difficulty is only mathematical, while the second aspect involves physical as well as mathematical difficulties.

A. THE ELECTRIC FIELD INSIDE CAVITIES IN A CONTINUOUS DIELECTRIC

Electrostatic field theory is treated in general text books[1] and with special reference to dielectric properties by Böttcher[2]. Fröhlich[3] uses field theory but is more concerned with the microscopic aspects of dielectric theory.

Electrostatic field theory is concerned with solving Laplace's equation for the electrostatic potential Φ

$$\nabla^2 \Phi = \frac{\partial^2 \Phi}{\partial x^2} + \frac{\partial^2 \Phi}{\partial y^2} + \frac{\partial^2 \Phi}{\partial x^2} = 0 \qquad (8.1)$$

for various configurations of dielectric or conducting bodies in space.

The solution of this equation is straightforward for a sphere in a uniform field. Assume a dielectric sphere of radius a and dielectric constant ε_2 in a medium of dielectric constant ε_1 which is assumed of infinite extent. In the outer medium at infinite distance from the sphere a uniform field E prevails, in the z direction. For this problem it is convenient to use as coordinates z and the distance r from the centre of the sphere and to express the potential in the form

$$\Phi(r, z) = -Ez + \Phi'(r, z) \qquad (8.2)$$

where $\Phi'(r, z)$ represents the disturbance due to the sphere and will be of different form inside the sphere and outside it. Although the potential is continuous at the surface of the sphere, its radial derivative is discontinuous, according to

$$\varepsilon_2 \left(\frac{\partial \Phi}{\partial r}\right)_2 = \varepsilon_1 \left(\frac{\partial \Phi}{\partial r}\right)_1. \tag{8.3}$$

In other words, the normal component of the electrical flux is continuous. The differential is taken in each case at the surface and the index 1 refers to the outside. In terms of E this means that the radial components at the boundary between the two media are related as

$$\varepsilon_2 E_{n2} = \varepsilon_1 E_{n1}. \tag{8.3a}$$

The discontinuity of the field can be interpreted as due to bound charges at the interface.

The solution of Laplace's equation gives for $\Phi'(r, z)$ outside the sphere

$$\Phi_1'(r, z) = \frac{\varepsilon_2 - \varepsilon_1}{2\varepsilon_1 + \varepsilon_2} \cdot \frac{a^2}{r^3} Ez. \tag{8.4}$$

This is the same potential that would follow for a dipole of infinitely small size, situated at the centre of the sphere and having a moment

$$m = \frac{\varepsilon_2 - \varepsilon_1}{2\varepsilon_1 + \varepsilon_2} a^3 E. \tag{8.5}$$

This means that the surface charges on the polarized sphere simulate a point dipole which is parallel to the field for $\varepsilon_2 > \varepsilon_1$ and anti-parallel to it for $\varepsilon_2 < \varepsilon_1$. This is characteristic for an isotropic sphere.

The contribution to the potential within the dielectric sphere is given by

$$\Phi_2'(r, z) = \frac{\varepsilon_2 - \varepsilon_1}{2\varepsilon_1 + \varepsilon_2} Ez \tag{8.6}$$

which defines a uniform field in the z direction. This potential is added to $-Ez$ according to equation (8.2), to give the potential within the dielectric sphere as

$$\Phi_2(r, z) = -\frac{3\varepsilon_1}{2\varepsilon_1 + \varepsilon_2} Ez. \tag{8.7}$$

This implies also a uniform field in the z direction. For $\varepsilon_2 < \varepsilon_1$ the field within the sphere is larger than the field in the external dielectric, being enhanced by the bound surface charges. For $\varepsilon_2 = 1$ the field in the empty spherical cavity within the dielectric may be called the cavity field

$$G = \frac{3\varepsilon_1}{2\varepsilon_1 + 1} E. \tag{8.8}$$

The magnitude of the electric field within a cavity in a dielectric depends on the shape of the cavity, but it is always $\geqslant E$. It is interesting to consider, still for a continuum, what happens to the field in a spherical cavity if this cavity is partially filled with dielectric material. In that case the field in the remaining empty portions increases.

An example of interest for the microscopic case can be constructed as follows. Assume a spherical cavity of radius a in a dielectric of dielectric constant ε. Now place into this cavity a concentric sphere of radius a_1 of the same dielectric. If $a - a_1 \ll a$ the inner sphere is surrounded by an infinitesimally thin shell of vacuum. This shell does not prevent the polarization of the inner sphere from being very nearly the same as for the bulk dielectric. Thus, the polarization/unit volume is given by

$$4\pi P = (\varepsilon - 1)E \qquad (8.9)$$

according to equation (2.10). Nevertheless, the inner sphere is surrounded by vacuum, and we may ask what magnitude E_i of uniform field would be needed to polarize it to the extent to which it is polarized. This field can be deduced from equation (8.5) by

$$P = m\frac{4\pi}{3}a^3 = \frac{3}{4\pi}\frac{\varepsilon - 1}{\varepsilon + 2}E_i. \qquad (8.10)$$

Comparison of equations (8.9) and (8.10) gives

$$E_i = \frac{\varepsilon + 2}{3} E \qquad (8.11)$$

for the effective field acting on the sphere in the cavity. The real field in the empty spherical shell which surrounds the sphere is of course not uniform.

B. THE STATIC DIELECTRIC CONSTANT FOR A REAL DIELECTRIC

Dielectric materials are not continuous, but made up of atoms which exert various forces on each other and move spontaneously. It may be convenient to distinguish three categories:

1. Materials which consist of distinguishable units, usually molecules, which may be polarized by a field but have no permanent moment in the absence of a field.
2. Materials with distinguishable units which have permanent dipole moments.
3. Solids where individual polarizable units cannot be sensibly distinguished and where the polarization has to be treated in terms[4] of the thermal vibrations of a crystal lattice (see Chapter 16, Section G).

Dielectric theory is largely concerned with the first two categories. The accepted approach due to Lorentz is to consider a spherical cavity in the dielectric, and to treat the material outside this cavity as a continuum while the particulate nature of the matter in the cavity is taken into account. The electric field within the cavity is in general[3]

$$\mathbf{E}_i = \mathbf{G} + \mathbf{R} \tag{8.12}$$

where $\mathbf{G}$ is the cavity field and $\mathbf{R}$ the reaction field due to the matter in the cavity. Vectorial quantities have to be used since the directions of the various field quantities, which coincide with that of E for a continuum, are no longer necessarily the same.

For the simplest model of a material of category 1 the field E_i in the vacuum within the cavity can be shown[3] to have the same value which has been deduced for a continuum by the somewhat artificial argument which leads to equation (8.11). This model contains N polarizable elements/unit volume in a cubic array, such that each element has an elastic (or optical) polarizability α_∞. The polarization/unit volume is now

$$P = Nm \tag{8.13}$$

$$= N\alpha_\infty E_i \tag{8.14}$$

where E_i is the field acting on an element. This field can be shown to be, in analogy to equation (8.11),

$$E_i = \frac{\varepsilon_s + 2}{3} E \tag{8.11a}$$

where ε_s is the static dielectric constant. Now, the polarization is quite generally given by equation (8.9). Elimination of E between equations (8.14) (using 8.11a) and (8.9) leads to the expression

$$\frac{\varepsilon_s - 1}{\varepsilon_s + 2} = \frac{4\pi}{3} N\alpha_\infty \tag{8.15}$$

which is known as the Clausius–Mossotti equation. For the model in question all fields (E_i, R, G and E) are always parallel to one another, as in the case of macroscopic dielectric spheres, and vector notation is superfluous. The Clausius–Mossotti equation is used in optics, but does not hold very accurately even for optical frequencies[2].

The Clausius–Mossotti equation has been used by Debye[5] for materials containing permanent dipoles, in the form (see equation 2.59)

$$\frac{\varepsilon_s - 1}{\varepsilon_s + 2} - \frac{\varepsilon_\infty - 1}{\varepsilon_\infty + 2} = \frac{N\mu^2}{3kT}. \tag{8.16}$$

8. ELECTROSTATIC INTERACTION

This equation is theoretically not admissible for reasons to be discussed, but it is useful because it gives an upper limit to the dielectric constant or dielectric loss which may be expected for a given material with a given concentration of dipoles. This equation is used in Appendix III to give a guide to the sensitivity of dielectric measurements. For gases or dilute solutions in liquids equation (8.16) can be disproved only if μ is accurately known. However, this equation breaks down completely when the concentration of dipoles in liquids becomes so large that ε_s should approach infinity. The Clausius–Mossotti catastrophe which signifies an increase of ε_s towards infinity with rising N or μ does not occur except in ferroelectrics, and these are often ionic materials where dipoles cannot be distinguished.

Equation (8.16) is inadequate because it takes no proper account of the way in which an applied field acts on a dipole. This inadequacy is particularly important for permanent dipoles which orient by rotation. It may be appreciated relatively simply in Onsager's[6] treatment of the dielectric constant of a dipolar liquid.

Onsager's theory concerns molecules with an isotropic polarizability α_∞ and a permanent dipole moment of value μ_0 in vacuo. The dipole is considered to be situated in the centre of a spherical cavity. This assumption excludes short-range forces due to an ordered array of neighbouring molecules. Such forces are generally present in solids and may also play a part in associated liquids where molecules form clusters.

The filling of the cavity is simplified in that not a representative sample of molecules is included but only one, in the centre of the cavity. This simplification is permissible[3] in view of the assumptions made.

The introduction of a dipole into the centre of the cavity complicates the calculation of the potentials Φ_2 and Φ_1 in and outside the cavity only slightly. Onsager shows that in statistical average over the coordinates of surrounding molecules we still have in the cavity a vectorial equation

$$\mathbf{E}_i = \mathbf{G} + \mathbf{R} = \frac{3\varepsilon_s}{2\varepsilon_s + 1}\mathbf{E} + \frac{2(\varepsilon_s - 1)}{(2\varepsilon_s + 1)a^3}\mathbf{m} \qquad (8.17)$$

formally similar to a scalar equation which may be deduced for a dielectric which obeys the Clausius–Mossotti equation. However, the vectorial quantity $\mathbf{m}$ is no longer always parallel to $\mathbf{E}$ because the direction of the dipole fluctuates spontaneously. The moment of the central dipole is, at a given instant, given by

$$\mathbf{m} = \mu\mathbf{u} + \alpha_\infty \mathbf{E}_i \qquad (8.18)$$

where $\mathbf{u}$ is a unit vector in the direction of the dipole. The scalar quantity α_∞ in this equation may be expressed by ε_∞ using the Clausius–Mossotti equation, which would hold if the molecules were not dipolar, but E_i contains a contribution from the dipole as well as from the polarizability. This means that the vectors **u**, **m**, **E** and $\mathbf{E}_i$ point in general in different directions.

The essential feature of Onsager's argument is that the field acts on the dipole by way of a force couple, and not an elastic distortion, as in the case of a molecule without permanent dipole moment. The force couple is the vectorial product of the dipole moment and the internal field $\mathbf{E}_i$. However, the reaction field **R** (see equation 8.17) is in the average always parallel to **m**. Hence one may calculate the effective force couple as an average over

$$\mathbf{T} = \mathbf{G} \times \mathbf{m} \tag{8.19}$$

and **R** does not contribute to the torque. As a consequence, the enhancement of the dielectric constant by electrostatic interaction is less than in the Clausius–Mossotti case.

Onsager's calculation leads to an equation for the static dielectric constant. For pure liquids whose volume equals the sum of the volumes of the molecules this may be written

$$\varepsilon_s - \varepsilon_\infty = \frac{3\varepsilon_s}{2\varepsilon_s + \varepsilon_\infty} \cdot \frac{4\pi\mu^2 N}{3kT} \left(\frac{\varepsilon_s + 2}{3} \right)^2. \tag{8.20}$$

The Onsager equation (8.20) agrees much better with experiment than the Clausius–Mossotti equation, in that ε_s remains finite for finite μ^2. This equation has been used to calculate dipole moments[2] for liquids, with results which agree reasonably well with moments deduced from measurements for gases. The fit may be further improved by the introduction of a parameter g, multiplied into μ^2. Kirkwood[7] and other authors have derived a constant parameter g on the basis of specific assumptions about the contents of the spherical cavity.

Fröhlich's[3] general treatment of a dipolar dielectric operates with a number of polarizable units of the same kind (say identical dipolar molecules) within a large spherical region. A given unit exhibits various dipole moments **m** in different directions which occur in the course of thermal fluctuations, with a certain probability. When a given moment **m** of the unit does occur, it causes a polarization **m*** within the spherical region as a whole. The value of **m*** is unequal to **m** because of the short range interactions between the polarizable units. As each unit fluctuates, the scalar product **mm*** fluctuates also. The average of this product over time, in unchanging conditions, gives the fluctuation of the dipole moment of the spherical region as

$$\overline{M^2} = N\overline{\mathbf{mm}^*}. \tag{8.21}$$

The magnitude of this fluctuation is related to the static dielectric constant. However, the relationship depends on whether the spherical region is embedded in vacuum, or in its own medium. This is so because a polarized sphere has an external field which reacts back on the polarizable units within the sphere. For a sphere in vacuo

$$\overline{M^2_{\text{vac}}} = \frac{9kTv}{4\pi} \cdot \frac{\varepsilon_s - 1}{\varepsilon_s + 2} \tag{8.22}$$

while for a sphere embedded in its own medium

$$\overline{M^2} = \frac{3kTv}{4\pi} \cdot \frac{(2\varepsilon_s + 1)(\varepsilon_s - 1)}{3\varepsilon_s}. \tag{8.23}$$

For $\varepsilon_s - 1 \ll 1$ both equations reduce to the same value

$$\overline{M^2} = \frac{3kTv}{4\pi} \cdot (\varepsilon_s - 1) \tag{8.23a}$$

which may be compared with equation (4.18) which has been derived disregarding electrostatic interaction. Equation (8.23a) and (4.18) agree except for a numerical factor 3, due to the bistable model being one-dimensional while equations (8.22) and (8.23) refer to three dimensional averages.

For $\varepsilon_s \to \infty$, equation (8.22) gives a constant value for $\overline{M^2}$. This limitation of the fluctuation is due to the external field of the sphere[3], a point which will be discussed in Section E of Chapter 16. For the sphere in its own medium the fluctuation tends to infinity as ε_s does.

The difference between equations (8.22) and (8.23) is due to macroscopic considerations, or, in other words, it is a consequence of electrostatic field theory. The more difficult aspect of Fröhlich's treatment is microscopic in that it concerns the derivation of the product $\overline{\mathbf{mm}^*}$ from molecular properties. Fröhlich's equation[3]

$$\varepsilon_s - \varepsilon_\infty = \frac{3\varepsilon_s}{2\varepsilon_s + \varepsilon_\infty} \cdot \frac{4\pi N}{3} \frac{\overline{\mathbf{mm}^*}}{kT} \tag{8.24}$$

provides a general solution for the static dielectric constant in terms of molecular properties.

C. RELAXATION IN THE PRESENCE OF ELECTROSTATIC INTERACTION

Any exact theory of relaxation in the presence of electrostatic interaction needs to be based on the theory of fluctuations. For that purpose it is convenient to operate with a sphere in vacuum.

A dielectric may in general contain several mechanisms of polarization. For instance a material may contain large molecules with permanent dipole moments which relax slowly (see Chapter 13), small molecular groups with permanent moment which relax rapidly, and a still more rapidly relaxing optical polarization. These mechanisms may be independent of each other on a molecular scale, although they need not be. If they are independent as far as short range forces are concerned they may still influence one another by way of their long range electrostatic interaction.

For a dielectric with a dipolar and an electronic mechanism Fröhlich[8] considers a spherical dielectric specimen and the contribution to the free energy from the two mechanisms. If the sphere is in vacuum, i.e. an external medium of $\varepsilon_1 = 1$, the free energy of polarization is given as a sum in which there are no cross terms from the fluctuations according to the two mechanisms. For $\varepsilon_1 > 1$ in the outer medium cross terms appear, that is, the mechanisms interact by way of the external field of the sphere. Scaife[9] shows, macroscopically, that more than two mechanisms will also not interact for a spherical specimen in a vacuum. In that case the total polarization is given by

$$\overline{M_{\text{vac}}^2} = 3kT \sum_1^n \alpha_j \tag{8.25}$$

where the α_j are polarizabilities for the different mechanisms. Scaife[10] bases on this argument a proposal for the display of relaxation data discussed in Chapter 7. The plot proposed is useful even though $\overline{M_{\text{vac}}^2}$ cannot be directly measured but has to be calculated from the measured $\varepsilon^*(\omega)$, and even though there is no certainty that different mechanisms are independent on a molecular scale.

The intricate distinction between macroscopic and microscopic interactions does not cause difficulties for a model dielectric which obeys the Clausius–Mossotti equation. It has already been shown in Chapter 2 for a Clausius–Mossotti model with a single relaxation time that equation (8.11a), quoted there without proof, leads to a simple relaxation equation. The single intrinsic relaxation time becomes a longer but still single extrinsic relaxation time, according to equation (2.65). This is due to the simplicity of the model, where external field, dipole moment and reaction field always point in the same direction so that a scalar differential equation can be used to describe relaxation. The polarizable units of the model behave like macroscopic spheres in a cubic array.

Conditions are more difficult for a dielectric obeying the Onsager model. A first tentative attempt to extend Onsager's equation (8.20) to dynamic conditions was made by Cole[11]. He simply put

$$G = \frac{3\varepsilon^*(\omega)}{2\varepsilon^*(\omega) + 1} \tag{8.26}$$

which is an *ad hoc* generalization of the cavity field. This attempt was not successful in that[12] it leads to a dependence of $\varepsilon^*(\omega)$ on frequency which cannot be represented by a super-position of linear differential equations (see equation 5.1). The expression also gives a maximum of $\varepsilon''(\omega)$ above the highest point of the Cole–Cole semi-circle which would mean, in physical terms, that resonance is present while Onsager's model contains no indication of resonance.

The difficulty of the problem in question may be seen when it is put in terms of a differential equation in analogy to equation (2.63). Onsager's argument may be expressed by stating that the field **G** is acting on the dipole since, in statistical average over the coordinates of surrounding molecules, the reaction field **R** of the dipole has no effect in turning it towards the field direction. We might, in analogy to equation (2.63), attempt to write a differential equation with **G** as driving field so that

$$\tau_0 \frac{d\mathbf{P}_D}{dt} + \mathbf{P}_D(t) = \alpha_D \mathbf{G} \qquad (8.27)$$

where τ_0 and α_D are constants, while $\mathbf{P}_D$ is the polarization of the dipole in the cavity. However, the connection between **G** and $\mathbf{P}_D$ is no longer clear in the dynamic case, since $\mathbf{P}_D$ and **G** both fluctuate and are unambiguously connected with one another only when averages over long times are taken. Equation (8.17) is no longer valid if the fields are defined for an instant of time while **m** fluctuates.

Powles[12] solves the problem by an artifice. He defines a time dependent field H, which reduces to **G** in the static case, so that

$$\frac{H}{E} = \frac{3\varepsilon_\infty}{2\varepsilon_\infty + 1} + \left(\frac{3\varepsilon_s}{2\varepsilon_s + 1} - \frac{3\varepsilon_\infty}{2\varepsilon_\infty + 1}\right) \frac{1}{1 + i\omega\tau}. \qquad (8.28)$$

Powles shows that for

$$\tau = \frac{3\varepsilon_s}{2\varepsilon_s + \varepsilon_\infty} \tau_0 \qquad (8.29)$$

equation (8.28) leads to a Debye equation

$$\varepsilon^*(\omega) - \varepsilon_\infty = \frac{\varepsilon_s - \varepsilon_\infty}{1 + i\omega\tau} \qquad (8.30)$$

with τ as relaxation time.

Powles' and Cole's formulae are not based on physical considerations. Glarum[13] approaches dielectric relaxation in a dielectric which behaves according to Onsager's model on the basis of a general statistical formalism developed by Kubo[14] for dissipative phenomena. Kubo relates the decay function of a macroscopic variable to a suitable correlation function of fluctuations in the system. This approach has been touched upon, in an elementary way, in Section C of Chapter 4, while decay functions were considered in Section B of Chapter 5.

Glarum considers first a spherical specimen which contains a dielectric obeying the Onsager model, assuming $\varepsilon_\infty = 1$. He defines a decay function $\Psi(t)$ which refers to the moment $M(t)$ of this sphere (see Section B, Chapter 5). Furthermore, he defines another decay function $\psi(t)$ which refers to the relaxation of an individual dipole. Glarum connects the macroscopic and microscopic decay functions by virtue of the fact that the fluctuation of a macroscopic body is due to a summation (with appropriate sign) of the fluctuations of its microscopic constituents. From the macroscopic decay function for the dielectric sphere the decay function for a plate capacitor may be deduced on the basis of field theory.

Finally, Glarum makes allowance for $\varepsilon_\infty \neq 1$ and defines quantities $\dot{\Psi}'(t)$ and $\dot{\psi}'(t)$ for the macroscopic and microscopic decay functions of dipolar mechanisms in a medium of given ε_∞. This step introduces some uncertainty. Glarum arrives at an expression for $\varepsilon^*(\omega)$ in terms of the Laplace transform of the time derivative of the microscopic $\dot{\psi}'(t)$. He finds

$$\frac{\varepsilon^*(\omega) - \varepsilon_\infty}{\varepsilon_s - \varepsilon_\infty} = \frac{(2\varepsilon_s + \varepsilon_\infty)L(-\dot{\psi}'(t))}{3\varepsilon_s - (\varepsilon_s - \varepsilon_\infty)L(-\dot{\psi}'(t))} \qquad (8.31)$$

where for a finite number of relaxation times entering into $\dot{\psi}'(t)$ the Laplace transform of that function is

$$L(-\dot{\psi}'(t)) = \sum_{j=1}^{n} \frac{a}{1 + i''\omega\tau_j} \qquad (8.32)$$

a sum of Debye expressions. For a single relaxation time all coefficients in (8.32) except $a_1 = 1$ and $\tau_1 = \tau$ vanish, and it is found that the complex $\varepsilon^*(\omega)$ in equation (8.31) has a single relaxation time given by equation (8.29). That is, the electrostatic interaction does not spread a single relaxation time into two or more components.

When the microscopic decay function contains more than one relaxation time the connection between the microscopic and macroscopic distribution functions is rather complicated. Glarum gives the result for two microscopic relaxation times, which are found to correspond to two macroscopic relaxation times. He also gives the result for the Cole–Cole distribution, which converts from one circular arc to another circular arc, and for the skewed arc, which converts into another skewed arc.

Scaife[15] derives a generalization of Onsager's equation by an approach which is in principle similar to that of Glarum[13]. He finds, in agreement with other investigators, that the intrinsic and extrinsic relaxation times are of the same order even if ε_s and ε_∞ are large. However, the implicit formula derived by Scaife implies that a single intrinsic relaxation time leads to a distribution of extrinsic relaxation times, a result which differs from that of Glarum.

Scaife[16] criticizes Glarum's calculation on the ground that Glarum's macroscopic decay function $\Psi(t)$ is not consistently defined. In equation (6) of reference 13, $\Psi(t)$ signifies the decay of the polarization of a sphere. In the paragraph following equation (7) of the same reference, $\Psi(t)$ signifies the decay function of the total polarization in the outer region surrounding this sphere. These two are not equal. Glarum's conclusion that a single relaxation time is not converted into a distribution of relaxation time as a consequence of electrostatic interaction is according to Scaife erroneous because of the inconsistency in the definition of $\Psi(t)$.

Another treatment by Zwanzig[17], based on Kubo's formalism[14], applies to a model of permanent dipoles in a cubic lattice in which dipoles rotate at random in the way discussed by Debye[5]. This author arrives at the conclusion that a single intrinsic relaxation time leads to a series of extrinsic or macroscopic relaxation times, as a consequence of electrostatic interaction.

Cole[18] treats a model where ε_∞ is attributed to harmonic displacements of a charge. He also gives a discussion of the statistical approach to dielectric relaxation in the presence of electrostatic interaction.

The Onsager equation applies to liquids which conform to a simple model. It will be seen in Chapter 11 that equation (8.29) is not an outstandingly good fit with the experimental data of liquids, even though it is an improvement on the Clausius–Mossotti treatment. As for solids, Chapter 16 will show that the so far most successful theoretical approaches to relaxation in the presence of strong electrostatic interaction proceed by way of thermodynamics and of lattice dynamics, rather than by way of the fluctuations of dipoles.

REFERENCES

1. J. A. Stratton (1941). "Electromagnetic Theory." McGraw Hill, New York.
2. C. J. F. Böttcher (1952). "Theory of Electric Polarisation." Elsevier, Amsterdam.
3. H. Fröhlich (1958). "Theory of Dielectrics." Oxford University Press, London.
4. B. Szigeti (1949). *Trans. Faraday Soc.*, **45**, 155.
5. P. Debye (1945). "Polar Molecules." Dover Publications, New York.
6. L. Onsager (1936). *J. Am. Chem. Soc.*, **58**, 1486.
7. J. G. Kirkwood (1939). *J. Chem. Phys.*, **7**, 911.
8. H. Frölich (1956). *Physica*, **22**, 898.
9. B. K. P. Scaife (1961). E.R.A. report Ref. L/T405.
10. B. K. P. Scaife (1962). E.R.A. report Ref. L/T416. Also (1963). *Proc. Phys. Soc.*, **81**, 124.
11. R. H. Cole (1938). *J. Chem. Phys.*, **6**, 385.
12. J. G. Powles (1953). *J. Chem. Phys.*, **21**, 633.
13. S. H. Glarum (1960). *J. Chem. Phys.*, **33**, 1371.

14. R. Kubo (1957). *J. Phys. Soc. Japan*, **12**, 570.
15. B. K. P. Scaife (1965). E.R.A. report Ref. N. 5132/1017.
16. B. K. P. Scaife. Personal Communication.
17. R. Zwanzig (1963). *J. Chem. Phys.*, **38**, 1766.
18. R. H. Cole (1965). *J. Chem. Phys.*, **42**, 637.

Note added in proof:

Fatuzzo and Mason[19] derive a generalization of the Onsager equation (8.20) which involves a time lag of the reaction field R. They find that a single intrinsic τ leads to more than one extrinsic τ. These authors claim that a modification of Glarum's[13] macroscopic decay function brings his treatment into agreement with their own[19] and Scaife's[15].

19. E. Fatuzzo and P. R. Mason (1967). *Proc. Phys. Soc.*, **90**, 729, 741.

CHAPTER 9

Resonance Absorption

The theories summarized so far refer mainly to the relaxation of the orientational polarization P_D, that is, to that part of the polarization which is due to the ordering of permanent dipoles. However, matter can take up energy out of a field even in the absence of permanent dipoles if the field perturbs oscillations of one kind or another. The polarization which is caused by this mechanism is called the electronic or optical polarization P_∞. It is an over-simplification to neglect the time dependence of P_∞, even though the assumption of a time independent P_∞ is a fair approximation in most practical dielectrics, at the frequencies below 10^{11} c/s or so which are accessible to electrical engineering.

A. MICROWAVE ABSORPTION IN GASES

Gases differ from solids and liquids in so far as marked effects of the time dependence of P_∞ may be observed at frequencies below 10^{11} c/s, i.e. at microwave frequencies. In condensed matter such effects may be observed optically in the infra-red region of the spectrum, but at "electrical" frequencies they are badly documented or non-existent. Besides, in the case of gases the optical polarization may be described as a property of individual molecules, while conditions in solids and liquids are less easy to visualize.

An individual atom or molecule, whether it has a permanent dipole moment or not, consists of negative and positive charges. In terms of classical physics the molecule may be considered as a harmonic oscillator. This means that the positive and negative charges have centres of gravity whose distance can be altered only by overcoming a restoring force. An oscillator of this kind takes up energy out of an electric field, at a resonant frequency, determined by the restoring force; the energy uptake in the absence of damping tends to infinity.

Classical physics is inadequate when dealing with individual molecules and the interaction of molecules and field is better described by quantum mechanics. Individual molecules have discrete energy states and they absorb energy out of an alternating field of frequency v so that

$$hv = \Delta W \qquad (9.1)$$

where ΔW is the energy difference between two quantum levels and h Planck's quantum. The molecule can take up or radiate energy only at the appropriate frequencies and in the appropriate amounts determined by (9.1). Each individual molecule is characterized by an optical spectrum with a number of discrete frequencies.

Optical spectra at frequencies accessible to electrical engineering correspond to very small energy differences as compared with the spectral lines of the visible or infra-red spectrum. The smallness of ΔW may be appreciated when it is compared with kT. For $v = 10^{10}$ c/s, ΔW is only $6 \times 10^{-3} \, kT$ at room temperature while the energy/degree of freedom is $\frac{1}{2}kT$ in equilibrium.

In spite of the smallness of the energy represented by microwave spectra, such spectra can be investigated in great detail for gases at low pressures. Microwave spectroscopy[1] provides a great deal of information about molecules and has important applications, but these need not be discussed here.

When molecules are not isolated but exist within a gas or other assembly they no longer have sharp spectra because they collide or otherwise interact with their neighbours. These interactions result in thermal equilibrium, and this implies an equilibrium distribution of energies among the constituent members of the assembly. It also implies that a given molecule retains its energy only for a certain average time τ_1. In a gas this life time is the shorter the higher the pressure.

When we wish to measure the value of an energy level in an assembly of molecules by means of some external probe, say microwaves, the result is unsharp according to the Heisenberg uncertainty principle. If the energy level in question has a lifetime τ_1, then it can only be determined with an uncertainty δW given by

$$\delta W \simeq \frac{h}{\tau_1} \tag{9.2}$$

where h is Planck's quantum. When τ_1 is short the uncertainty of the energy determination may be large compared with the magnitude of the energy difference to be determined. That is we may have $\delta W \gg \Delta W$. In that case the magnitude of ΔW can not be determined, and spectroscopy is impossible. For gases at low pressures the life times τ_1 of the rotational and similar states of molecules are long because collisions are rare, and microwave spectroscopy is possible in spite of the smallness of ΔW. For pressures above about 10^{-1} mm Hg the microwave spectra of molecules become unsharp, and for condensed phases they are in general blurred beyond recognition.

B. THE BROADENING OF SPECTRAL LINES IN GASES

The broadening of line spectra implies a dissipation of energy. The line shape and energy dissipation may be calculated using statistical thermodynamics on the basis of either classical physics or quantum mechanics. A

9. RESONANCE ABSORPTION

calculation of this kind was first carried out for a gas by Lorentz[2]. He assumed that an atom or molecule is equivalent to a classical oscillator, which obeys an equation

$$m\left(\frac{d^2x}{dt^2} + \omega_0^2 x\right) = eE_0 e^{i\omega t}. \tag{9.3}$$

Here e and m are charge and mass respectively, while the right-hand side of the equation represents a periodic electrical field; ω_0 represents a restoring force. Lorentz assumed that the classical oscillators in question interchanged energy by collisions and deduced a theory of line broadening which is still a good approximation for the case that neither ω nor ω_0 is very small and that line broadening is not too great. For that case, which covers most typical spectra, Lorentz results may in close approximation be expressed by introducing a damping term proportional to dx/dt in the differential equation of an oscillator. Equation (9.3) is written for an individual oscillator, but the differential equation for the assembly as a whole may be formulated in terms of the polarization/unit volume as

$$\frac{d^2 P_\infty(t)}{dt^2} + \gamma \frac{dP_\infty(t)}{dt} + P_\infty(t) = \frac{\omega_0^2 \Delta\varepsilon}{4\pi} E(t) \tag{9.4}$$

where $P_\infty(t)$ is the polarization attributable to a certain mode of oscillation of characteristic frequency ω_0 which contributes a term $\Delta\varepsilon$ to the static dielectric constant, γ is a constant and $E(t)$ the electric field.

Equation (9.4) converts to the relaxation equation if we divide it through by ω_0^2, neglect the second order term and put

$$\frac{\gamma}{\omega_0^2} = \tau. \tag{9.5}$$

It is analogous to the differential equation of a circuit of capacitance, inductance and resistance, all in series (equation 1.46). Equation (9.4) may be integrated by introducing a complex dielectric constant ε^* so that

$$\varepsilon^*(\omega) - \varepsilon_\infty = \frac{\Delta\varepsilon}{1 - \left(\frac{\omega}{\omega_0}\right)^2 + i\left(\frac{\omega}{\omega_0}\right)\left(\frac{\gamma}{\omega_0}\right)} \tag{9.6}$$

represents the relationship between field and polarization for periodic fields. The solution for a step-function is given in Chapter 1.

Equation (9.6) is not identical with the dielectric relationship resulting from Lorentz' original treatment, which is discussed, for instance, by Gross[3]. However equations (9.4) and (9.6) are commonly used in the treatment of optical spectra, and represent the simplest formal description of resonance absorption. The behaviour of the complex dielectric constant as a function of frequency is quite different for resonance and relaxation, as may be seen from Figs 1 and 2. The resonance behaviour is sometimes described as

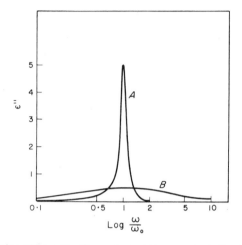

FIG. 1. ε'' as a function of log ω/ω_0. Curve A resonance case according to equation (9.6) with $\Delta\varepsilon = 1$, $\gamma/\omega_0 = 5$. Curve B, relaxation with $\Delta\varepsilon = 1$. After Herzfeld and Litovitz (1959). "Absorption and Dispersion of Ultrasonic Waves." (Academic Press, New York and London.)

anomalous dispersion. The constants in the illustrated case are chosen so that the resonance line is moderately damped. The characteristic feature of the resonance case is that $\varepsilon'(\omega)$ rises above the low frequency value and dips

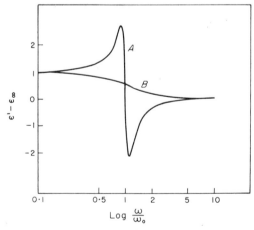

FIG. 2. $\varepsilon' - \varepsilon_\infty$ as a function of log $\dfrac{\omega}{\omega_0}$, for the same conditions as in (9.1).

9. RESONANCE ABSORPTION

to negative values, while the peak of $\varepsilon''(\omega)$ is much sharper than for relaxation. Equation (9.6) is often written with different arrangements of its constants. A comparison of optical and dielectric notations is given in Section D of this chapter.

Later authors modified the theory of Lorentz in various ways. Two important modern treatments of the broadening of spectral lines are due to van Vleck and Weisskopf[4] and independently to Fröhlich[5]. They lead to the same result although the models used are different. Van Vleck and Weisskopf treat collision broadening in a gas, while Fröhlich uses a generalized model which might apply also to an idealized solid. We shall here discuss mainly the van Vleck–Weisskopf treatment, since Fröhlich's method is included in his book on dielectric theory[5] and is thus more accessible.

Van Vleck and Weisskopf analyse Lorentz' treatment and criticize a feature which may apply equally to a classical or quantum mechanical model. This feature is the way in which the collisions between molecules are assumed to act. Van Vleck and Weisskopf assume "strong" collisions, in the sense that after a collision a molecule has no memory of its state before the collision. That is, a collision produces a random orientation or energy state, independent of the state before the collision. Lorentz assumes that states after the collision will be symmetrically distributed about the state before the collision, i.e. he does not assume complete randomness (see Section C of Chapter 4).

The interaction of the oscillators by strong collisions implies that, as a consequence of collisions, their energy states obey Boltzmann statistics; this is Fröhlich's assumption also. Both treatments lead to a result which may be expressed, according to Fröhlich, by

$$\varepsilon^*(\omega) - \varepsilon_\infty = \tfrac{1}{2}\Delta\varepsilon\left(\frac{1 - i\omega_0\tau}{1 - i(\omega_0 + \omega)\tau} + \frac{1 + i\omega_0\tau}{1 + i(\omega_0 - \omega)\tau}\right) \quad (9.7)$$

where ω_0 is a characteristic frequency, τ a time constant and $\Delta\varepsilon$ that part of the dielectric constant which corresponds to the polarization of the oscillators in question.

Fröhlich shows that equation (9.6) is equivalent to a decay function of the form

$$\Psi(t) = q\, e^{-\frac{t}{\tau}} \cos(\omega_0 t + \phi) \quad (9.8)$$

where q and ϕ are constants, determined by

$$\Delta\varepsilon = q\tau \cos\phi \quad (9.9)$$

and

$$\omega_0 \tau = -\tan\phi. \quad (9.10)$$

This represents damped oscillations with frequency $\omega_0/2\pi$ about the decaying trend of the polarization, and is more simple than the decay function which follows from equation (9.4) and is discussed in Chapter 1.

Van Vleck and Weisskopf discuss the connection between resonance absorption and Debye relaxation. This is clear enough in terms of the differential equation (9.4), where the second order term becomes negligible against the first order term as damping increases. It is not obvious for all theories of line broadening, but van Vleck and Weisskopf[4] show that their treatment leads to a smooth transition from resonance to relaxation. The subject is reviewed in simple terms by Gross[3], and will be treated in more detail in Chapter 10.

A number of other theories have been advanced about line broadening in gases, based on different assumptions about the oscillators and the nature of the collisions between them; the subject is reviewed by Illinger[6]. All these theories have in common that they lead to a linear relationship between polarization and field. In terms of differential equations, the result of each theory may be represented by one or more linear second order differential equations. Alternatively an analogous *RCL* circuit may be devised. Linearity means that a complex dielectric constant may be defined which obeys the Kramers–Kronig relations (equations 5.20 and 5.21).

C. RESONANCE ABSORPTION IN SOLIDS

The absorption spectra of crystalline solids are at short wavelengths due to electronic transitions within atoms or molecules and at longer wavelengths due to vibrations of the crystal as a whole. The dynamic theory of crystal lattices is treated in an authoritative manner by Born and Huang[7], and will here only be mentioned very briefly. The lattice vibrations fall into two categories: optical and acoustic. The latter vibrations do not involve a change of the polarization and do not interact with electromagnetic fields. The optical vibrations correspond to oscillatory displacements of charges. Optical and acoustic modes together constitute the thermal vibrations whereby the atoms or ions, etc. within the crystal achieve thermal equilibrium.

The most important optical vibration in a simple ionic crystal like KCl may be visualized as a movement of K and Cl ions in opposition to each other, as in the case of a dipole whose length is oscillating. The frequency of this vibration in the alkali halides is of the order 10^{13} sec^{-1}. The infra-red spectra which correspond to the vibration in question have been studied extensively, since they give information about cohesive forces and the geometry of the electron clouds.

The broadening of the infra-red spectra of crystalline solids is in general a

9. RESONANCE ABSORPTION

difficult subject. However, in first approximation it may be treated quite simply. Born and Huang state that "for the purpose of analysing the empirical data in the neighbourhood of ω_0 it proves convenient to use a dispersion formula which takes account of the energy dissipation in an *ad hoc* way". Born and Huang give for the purpose a differential equation which is identical with equation (9.4) except for the significance of the constants. In a solid the polarizability expressed by $\Delta\varepsilon$ and the characteristic frequency ω_0 belong to a lattice vibration, while in a gas they are attributes of an atom or molecule. However, as long as we are not concerned with the physical significance of the constants and line shapes cannot be measured with great accuracy, equation (9.4) appears to be adequate for both gases and solids.

D. THE DIELECTRIC AND OPTICAL CONSTANTS

Since experimental evidence on resonance absorption is often obtained by optical methods it will be useful to correlate the optical and dielectric terminologies.

The optical constants refer to the velocity and damping of a plane electromagnetic wave in a homogeneous medium. For such a wave the amplitude as a function of distance and time is given by

$$A = A_0 \, e^{i\omega\left(\frac{xn^*}{c} - t\right)} \tag{9.11}$$

where A_0 is a constant, c the velocity of light in vacuo and n^* the (complex) refractive index. For materials of unit magnetic permeability the refractive index is equal to the square root of the complex dielectric constant. The complex refractive index is defined in various ways by different authors, as

$$n^* = n(1 + ik) \tag{9.12a}$$

$$n^* = n + ik \tag{9.12b}$$

or

$$n^* = n - ik \tag{9.12c}$$

where n and k are real constants. We shall here use the first definition, in accordance with Born and Huang. With definition (9.12a), the real and imaginary parts of the dielectric and optical constants respectively are connected by the equations

$$\varepsilon' = n^2(1 - k^2) \tag{9.13}$$

$$\varepsilon'' = -2n^2 k. \tag{9.14}$$

Experimentally the optical constants may be determined by transmitting a light beam through a layer of dielectric and by measuring its refractive index n and its decrease in energy as a function of thickness. Alternatively, the beam may be reflected from a clean surface and the reflected fraction R of its energy measured. It may be shown that

$$R = \frac{(n-1)^2 + n^2k^2}{(n+1)^2 + n^2k^2} \tag{9.15}$$

where the thickness of the specimen is assumed large compared to the wavelength of light. Equation (9.15) implies that for high values of n the reflective

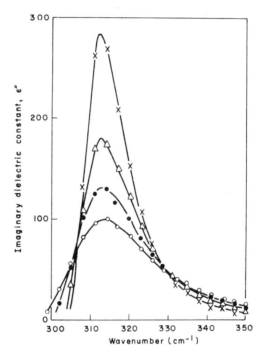

FIG. 3. ε'' as a function of temperature and wave number in LiF, near dispersion peak. ×—135°K; △—210°K; ●—300°K; ○—355°K (Gottlieb[9]).

power goes towards unity and nearly all the light is reflected. This consideration is important not only in optics but also for dielectric measurements at microwave frequencies on materials of high dielectric constant.

E. THE TRANSITION FROM RESONANCE TO RELAXATION

The transition from resonant to non-resonant absorption has been studied in gases, even though workers in the field of microwave spectroscopy are mostly interested in sharp spectra. However, theories of line broadening have

9. RESONANCE ABSORPTION

been compared with experiment and in some cases the damping was increased until the resonant character of an absorption line was lost. For instance, Bleaney and Loubser[8] measured absorption in gaseous ammonia due to the spectral line characteristic for inversion of the molecule. The frequency of this line is 2.33×10^{10} c/s (0.78 cm^{-1}) for very low pressures. For increasing pressure the line becomes wider and less distinct, while its position shifts to lower frequencies. This investigation gave good agreement with equation (9.7), but the theory does not explain all the observed facts.

While it is possible to find resonance absorption at electrical frequencies in gases of moderate pressure, and a gradual transition from resonance to relaxation is well documented (see also Chapter 10), resonance is difficult to discern in solids and liquids at electrical frequencies.

The transition from resonance to relaxation has been approached also for solids, by taking a series of measurements at frequencies and temperatures where $h\nu$ and kT are comparable. The evidence is conclusive for the alkali halides, for frequencies in the far infra-red. Figure 3 shows the dependence of ε'' on frequency for LiF. It may be seen[9] that the line broadens very appreciably for 355°C, although it is still quite distinct; for the frequency in question $h\nu = kT$ at 45°C. The line shifts only little with temperature (very much less than a Debye peak would shift). Gottlieb uses equation (9.6) for his calculations.

For frequencies appreciably lower than 10^{13} c/s optical measurements become increasingly difficult, and very complex techniques have to be used[10]. Nevertheless optical measurements have been extended to frequencies below 3×10^{11} c/s, wavelengths above 1 mm. However, the accuracy of measurement is relatively low, particularly for older work. Palik[11] gives a bibliography of the far infra-red spectral region from 1892–1960. Recent work in the far infra-red, achieving increasingly good resolution, has been concerned with alkali halides, long-chain molecules[12] and semi-conductors[13].

For a wavelength of about 1 mm the optical techniques of the far infra-red overlap with electrical microwave techniques. Genzel et al.[14] made measurements on alkali halides which link the two regions. Figure 4 gives n and k (based on definition 9.12c) for NaCl. The increase of n with decreasing wavelength below about 0·1 mm indicates indubitable resonance absorption. The increase of k from about 10 mm onwards is in this case also clearly a resonance phenomenon, since it links up with an infra-red maximum. However, in the absence of infra-red data it would be difficult to prove that the increase of loss with increasing frequency is not due to a relaxation but a resonance.

In general it is not safe to assume that the quantity $\varepsilon_s - \varepsilon_\infty$ measured for a material is exclusively due to the orientation of dipoles and to conduction phenomena. A part of this quantity may be due to over-damped oscillations with ω_0 in the electrical frequency region. This ambiguity of interpretation is not important for dielectrics with large dipoles where the temperature dependence of $\varepsilon_s - \varepsilon_\infty$ is easily measured. However, over-damped resonance has to be borne in mind as a possible mechanism in not very polar dielectrics.

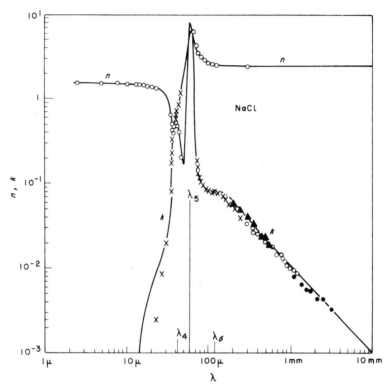

Fig. 4. n and $n^2 K$ for NaCl as a function of wavelength (Genzel[14] et al.).

REFERENCES

1. J. Sheridan (1962). "Microwave Spectroscopy in Gases, Progress in Dielectrics", Vol. 4, p. 1. (Birks and Hart, eds.) Heywood, London.
2. H. A. Lorentz (1915). "The Theory of Electrons." Dover Publications, New York.
3. E. P. Gross (1955). *Phys. Rev.*, **97**, 395.
4. J. H. van Vleck and V. F. Weisskopf (1945). *Rev. Mod. Phys.*, **17**, 227.
5. H. Fröhlich (1958). "Theory of Dielectrics." Oxford University Press, London.
6. K. H. Illinger (1962). "Dispersion and Absorption of Microwaves in Gases and Liquids, Progress in Dielectrics", Vol. 4, p. 37.
7. M. Born and K. Huang (1954). "Dynamic Theory of Crystal Lattices." Oxford University Press, London.
8. B. Bleaney and J. H. N. Loubser (1950). *Proc. Phys. Soc.* A, **63**, 483.
9. M. Gottlieb (1960). *J. Opt. Soc. Am.*, **50**, 343.

10. G. O. Jones, D. H. Martin *et al.* (1961). *Proc. R. Soc.* A, **260**, 510 (1960) and ibid **261**, 10.
11. E. D. Palik (1960). *J. Opt. Soc. Am.*, **50**, 1329.
12. S. B. Field and D. H. Martin (1961). *Proc. Phys. Soc.*, **78**, 625.
13. Picus, Burstein, Henvis and Hass (1959). *J. Phys. Chem. Solids*, **8**, 282.
14. L. Genzel, H. Happ and R. Weber (1959). *Z. Phys.*, **13**, 154.

CHAPTER 10

Dielectric Relaxation in Gases

Gases are interesting from the point of view of the theory of relaxation because their dielectric relaxation times can in favourable cases be derived from kinetic theory. Unfortunately, relatively little experimental evidence is available on Debye type dielectric relaxation in gases, as distinct from the broadening of microwave spectra. This is not surprising because the measurement of dielectric loss in a gas far from resonance conditions demands the detection of a very small dissipation of energy/volume: gases have a low density unless very highly compressed or near their critical point, and the energy absorbed/molecule out of an electric field in a dielectric relaxation process is always small compared with kT.

Relaxation in gases as distinct from resonance may be studied more easily by ultrasonic methods, where relatively large energies may be stored/unit volume, as compared with the dielectric case. Much work has been done on the absorption and dissipation of ultrasonic waves in gases, and Herzfeld and Litovitz[1] show that a great deal of information can be derived from this work on the dynamic properties of molecules and the ways in which collisions between molecules cause damping and viscosity.

The present chapter examines the dielectric evidence on relaxation in so far as this can be separated from the more abundant work on the broadening of optical spectral lines, and attempts to interpret it in terms of the current theories of energy transfer by collisions. This subject has recently been reviewed by Illinger[2].

The experimental work on dielectric relaxation in gases has mostly been carried out in the highest frequency region (10^9–10^{11} c/s) of the electrical spectrum. In this region cavity resonators may be used as these lend themselves to use at varying pressure, and so to work on gases. Experimental methods have been described for use up to pressures of over 1000 atmospheres[3,4]. The cavity resonators in question operate only over a narrow range of frequencies, so that measurements need to be made at spot frequencies, while pressure and temperature may be varied continuously.

The frequency region around 10^{10} c/s contains resonance lines for many gases. The broadening of these lines is clearly a special case of damping by collisions. However, the most straightforward cases of relaxation are those where the energy absorption out of the field is by orientation of permanent dipoles rather than by resonance.

A relaxation time τ may be ascribed both to resonant and non-resonant absorption. In the last chapter, the damping of resonance lines was in equation (9.4) characterized by a damping constant γ and in equation (9.7) by τ. A third characterization is often given by the half-width of a resonance line $\Delta \nu$.

These quantities are related by

$$\frac{1}{\tau} = \frac{\omega_0^2}{\gamma} = 2\pi \, \Delta\nu. \qquad (10.1)$$

The relaxation time for a resonance line may be derived from its width. For strong damping also the frequency of the line is affected by τ as shown in reference 8 of Chapter 9.

For a non-resonant peak the frequency of maximum loss is given by the Debye relations in first approximation, so that the relaxation time may be derived from the position of the peak. Non-resonant peaks have been investigated in detail for a series of substituted methanes, CH_3F, $CClF_3$, etc., which are called "symmetrical top" molecules because they are symmetrical about an axis of rotation. These molecules have more or less spherical shapes and often strong dipole moments. They are also of particular interest in the liquid and solid state. The symmetrical top molecule NH_3 has a famous microwave resonance line corresponding to the molecule turning inside out. This type of transition does not occur in the substituted methanes in question but they have other resonance lines in the relevant region between 10^9 and 10^{11} c/s. However, these lines are not very prominent in comparison with the non-resonant peak.

Birnbaum, Maryott and collaborators[5,6,7] have investigated non-resonant absorption in substituted methanes in the pressure region below and around one atmosphere. At those pressures the relaxation time τ is found to be proportional to the pressure

$$\tau = \text{const.} \, p \qquad (10.2)$$

and a Debye peak may be mapped as a function of p for constant frequency. Birnbaum[5] shows that the Debye equation transforms into

$$\frac{p^2}{\tan \delta} = a + bp^2 \qquad (10.3)$$

where a and b are constants related to τ and $\varepsilon_s - \varepsilon_\infty$ by the equations

$$\tau = \frac{1}{\omega p} \sqrt{\frac{a}{b}} \qquad (10.4)$$

$$\left(\frac{\tan \delta}{p}\right)_{\max} = \frac{1}{2\sqrt{ab}} \qquad (10.4b)$$

where equations (10.4) refer to the loss angle peak and ω is the (constant) frequency at which the measurements were made. Figure 1 shows that equation (10.3) holds well at pressures around an atmosphere, but that at lower pressures the absorption is higher than follows from a Debye curve. This is to be expected if resonance is present.

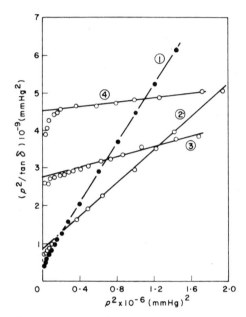

FIG. 1. Variation of $\dfrac{p^2}{\tan \delta}$ with p^2 for CH_3Br (Birnbaum[5], 1957). 1. 1193 Mc, 26°C; 2. 2216 Mc, 26·5°C; 3. 9280 Mc, 26°C; 4. 23,340 Mc, 24°C.

Figure 2 shows relaxation peaks for different temperatures, corresponding to linear plots according to equation (10.3). The magnitude of these peaks is strongly temperature dependent, while their position is relatively little dependent on temperature, a situation which is different in liquids and solids.

The non-resonant absorption spectra for CH_3F_1, CH_3Cl, CH_3Br, CH_3CN and CHF_3 were found to be wider than Debye peaks[2]. They could be represented by a Cole–Cole distribution of relaxation times. For CH_3Cl this is quoted as characterized by $\alpha = 0.051$, a narrow distribution. The gases $CClF_3$ and $(CH_3)_3N$ showed pure Debye peaks, within the limits of experimental error.

Maryott and Birnbaum[5,6] base the interpretation of their results on the van Vleck–Weisskopf[8] theory of the collision broadening of spectral lines. This theory includes the case of non-resonant absorption by the orientation of permanent dipoles as a limiting case, formally expressed by $\omega_0 = 0$. The result of the theory is an equation identical with the Debye equation, except that the effective dipole moment may be different from the static dipole moment of a molecule. The difference is due to the fact that in the case of strong collisions, assumed by van Vleck and Weisskopf, the rotational energy of a molecule may change as a consequence of collisions. Debye's original treatment assumes weak collisions, i.e. collisions or other interactions which perturb the condition of dipoles only slightly.

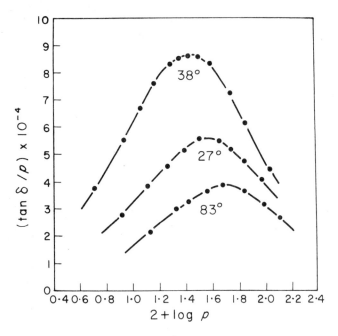

Fig. 2. Representative data for CH_3F. Plots of $\dfrac{\tan \delta}{p}$ as a function of log p at several temperatures (from Maryott et al.[6]).

The assumption of strong collisions on which the van Vleck–Weisskopf treatment is based implies that the relaxation time is of the order of τ_c, the average time which elapses between successive collisions. This follows because a strong collision is by definition energetic enough to destroy every after effect of the previous history of a molecule (see also Section D of Chapter 4).

The theory of van Vleck and Weisskopf is essentially an approach to the calculation of absolute values of dielectric relaxation times on the basis of kinetic gas theory. However, these authors do not claim that their treatment

10. DIELECTRIC RELAXATION IN GASES

is more than approximate. The difficulties of the theory are discussed, for instance, by Anderson[9] and Gross[10].

The simplest model for the kinetic theory of gases is an ideal gas consisting of rigid spherical atoms. For such a gas the interval τ_c between collisions can be calculated as a function of pressure and temperature on the assumption that all collisions are quite elastic. For such a gas the viscosity η is also fully determined, and may be written[1]

$$\eta = p\tau_{tr} \qquad (10.5)$$

where τ_{tr} is the time necessary for the adjustment of translational energies. This time is related to the time interval between collisions by

$$\tau_{tr} = 1\cdot 271\tau_c. \qquad (10.6)$$

For nearly ideal gases at 0°C and one atmosphere τ_{tr} is of the order 10^{-10} seconds.

Most molecules do not in fact behave as rigid spheres, and they exchange during collisions not only translational but also vibrational and rotational energies. In consequence, the time which elapses between collisions is not a unique measure of the relaxation time of all processes which occur in a gas. The complexity of the situation emerges most clearly when one considers ultrasonic relaxation because it is found that the relaxation time needed to adjust the energy content of a gas to a change of pressure is many times longer than τ_c. Herzfeld and Litovitz define a number Z, such that

$$\tau_u = Z\tau_c \qquad (10.7)$$

where τ_u is the ultrasonic relaxation time. This may briefly be defined as follows. The molecules of a gas have external degrees of freedom such as translational ones, and internal degrees of freedom, such as vibrational ones. When a gas in equilibrium with a thermal reservoir at temperature T is compressed it acquires excess energy. The external degrees of freedom respond very rapidly, reaching a temperature T_{tr} in negligible time. The internal degrees of freedom lag behind according to a differential equation

$$W + \tau_u \frac{dW}{dt} = \tau_u W(T_{tr}) \qquad (10.8)$$

where W is the momentary value of the energy of the internal degrees of freedom and $W(T_{tr})$ is that energy of the internal degrees of freedom which is in equilibrium with the temperature of the external degree of freedom. The relaxation time τ_u is therefore characteristic for the adjustment of the vibrational, etc. degrees of freedom. The differential equation (10.8) describes the relaxation phenomenon responsible for "excess" absorption in a gas beyond the absorption for an ideal gas of rigid spheres.

Table I shows a comparison of dielectric and ultrasonic relaxation times for some halogenated methanes and related molecules at one atmosphere. The dielectric data (τ_D) are deduced from references 5 and 6, while the acoustic ones are taken from Table 54.2, reference 1. The fourth column gives the Cole–Cole parameter α for the dielectric case.

TABLE I
Dielectric and ultrasonic relaxation times in gases at a pressure of 1 atmosphere

Gas	Temperature °C	τ_D sec $\times 10^{10}$	Cole–Cole parameter α	τ_u sec $\times 10^8$	$\tau_c \times 10^{10}$	Reference
CH_4	15			48	1·00	1
	109			107	1·27	
CH_3F	26	0·47	>0			5,6
	83	0·64				6
	100			320		1
CH_3Cl	26	0·47	0·05	20		5
	100			18		1
CH_3Br	26	0·56	>0			5
	30			7·5		1
CH_3I	26	0·60	>0	4·5		5
	100					1
CHF_3	26	1·03		48		5,6
	89	1·38				6
	100			42		1
$CClF_3$	28	1·34	0	24		1
	81	1·63				
CH_3CN	26	1·53				5
$(CH_3)_3N$	27	0·67	0			5

The table shows that the ultrasonic relaxation times are longer by a factor of the order 1000 or 10,000 than the dielectric ones. Comparison of the dielectric relaxation times with τ_c for the non-polar but closely related molecule CH_4 shows that τ_D is of the order of τ_c, that is the time between collisions. This agrees with the assumption of strong collisions—at any rate for the nearly spherical molecules in question. Since the gases in question are nearly ideal at one atmosphere, equations (10.5) and (10.6) linking the viscosity with τ_c are fairly well obeyed. This means that there is a close link between the dielectric relaxation time, the viscosity, and the relaxation time τ_{tr} for the translational degrees of freedom, while the ultrasonic τ_u refers to a process of another kind.

The great difference between the ultrasonic and dielectric relaxation times provides a fundamental and simple reason for the existence of a distribution of relaxation times. This reason, which was touched upon by van Vleck and

10. DIELECTRIC RELAXATION IN GASES

Weisskopf at a time when suitable ultrasonic data were not yet available, is as follows. Molecules which exchange their internal energies slowly remain after a collision in an internally unadjusted state for a time which is long compared with τ_c or τ_D. If a snapshot were taken during a time interval τ_D the molecules in the picture could be separated into distinguishable groups in varying stages of the adjustment of internal degrees of freedom. If these groups had, for instance, different diameters they would probably have different relaxation times. In other words, the different relaxation times of a distribution might refer to molecules in different stages of adjustment.

If a molecule takes part simultaneously in two relaxation processes one of which is fast compared with the other, then there is good reason to expect a more or less appreciable distribution of the fast relaxation time. It is interesting to note that Cole–Cole distributions with $\alpha \neq 0$ have been found for some of the symmetrical top molecules. The presence of a slow process does of course not exclude a single relaxation time for the fast one.

The table shows that τ_D increases with increasing temperature, while the ultrasonic relaxation time depends on temperature in an irregular way. Herzfeld and Litovitz[1] state that for an ideal monatomic gas the collision time as a function of temperature may often be represented by

$$\frac{\tau_c}{p} = T^{\frac{1}{2}} \cdot \frac{T}{T+C} \tag{10.9}$$

where C is a constant for a given gas. The increase of τ_c with increasing temperature, at constant pressure, is due to thermal expansion.

The dielectric relaxation time τ_D for the substituted methanes increases more strongly with temperature than corresponds to (10.9). The theoretical understanding of the temperature dependence of τ_D, as well as the calculation of the magnitude of τ_D, demands a detailed knowledge of molecular properties and their effect on collisions.

Anderson[9] gives a theory of collision broadening of spectral lines more detailed than that of van Vleck and Weisskopf[8] in that he considers that a collision need not be adiabatic but may change quantum states of the colliding molecules. This theory predicts that

$$\frac{\tau_D}{p} = \text{const.} \, T^{-m} \tag{10.10}$$

where m is restricted to values between 1 and 0·5. This formula is successful for NH_3, but not so successful for other symmetrical top molecules for which m is found[5] to be 1·59 for CHF_3, 1·60 for CH_3F and 1·27 for $CClF_3$. Birnbaum and Maryott[7] ascribe this discrepancy to the large dipole moments of the molecules in question. They suggest that during a collision the molecules exert a powerful torque on each other by the electrostatic effect of their differently oriented dipoles. The argument leads to a formula connecting the dipole moment of a molecule, its average angular momentum and the relaxation time. This fits in well with the data of Table I and explains the observed temperature dependence of the relaxation time of the molecules with large dipole moments.

Literature references up to 1961 on relaxation in gases are to be found in Illinger's review[2]. More recent work is concerned with theoretical refinements such as[11] transient dipoles introduced by collisions. Frenkel[12] gives a phenomenological treatment of dielectric relaxation for symmetrical top gases and binary gas mixtures which gives reasonable agreement between experimental and calculated collision frequencies and cross sections.

The discussion so far refers to conditions where the duration of collisions may be assumed small compared with the time between collisions. This assumption breaks down near the critical point. The relaxation behaviour of gases at pressures approaching the critical point has so far only been surveyed in a preliminary way. Phillips[3] measures ε' and ε'' as a function of pressure and temperature, for $CHClF_2$ and N_2O, as well as for mixtures of $CHClF_2$ with nitrogen. The experiments were carried out at a single frequency of 2.46×10^{10} c/s. Since equation (10.2) no longer holds for the high pressures used (up to 1000 atmospheres) the mapping of Debye peaks as a function of pressure is no longer possible.

It was estimated, however, that both for the gases at low pressure and for the liquefied N_2O the relaxation times were of the order 10^{-12} seconds. That is, while the dielectric relaxation time of the ideal gas increases with pressure, an increase of pressure beyond the critical point shortens the relaxation time to the value characteristic for a liquid.

REFERENCES

1. K. F. Herzfeld and T. A. Litovitz (1959). "Absorption and Dispersion of Ultrasonic Waves." Academic Press, New York and London.
2. K. H. Illinger (1962). "Progress in Dielectrics", Vol. 4, p. 37–101. Academic Press, London and New York.
3. C. S. E. Phillips (1955). *J. Chem. Phys.*, **23**, 2388.
4. M. G. Vallauri and P. W. Forsbergh (1957). *Rev. scient. Instrum.*, **28**, 198.
5. G. Birnbaum (1957). *J. Chem. Phys.*, **27**, 360.
6. A. A. Maryott, A. Estin and G. Birnbaum (1960). *J. Chem. Phys.*, **32**, 1501.
7. G. Birnbaum and A. A. Maryott (1958) *J. Chem. Phys.*, **29**, 1422.
8. J. A. van Vleck and V. Weisskopf (1945). *Rev. mod. Phys.*, **17**, 227.
9. P. W. Anderson (1949). *Phys. Rev.*, **76**, 647.
10. E. P. Gross (1955). *Phys. Rev.*, **97**, 395.
11. G. Birnbaum and A. Maryott (1962). *J. Chem. Phys.*, **36**, 2026.
12. L. Frenkel (1966). *J. Chem. Phys.*, **45**, 416.

CHAPTER 11

Relaxation in Liquids

The last chapter has shown that dielectric relaxation in gases at low pressures is difficult to observe experimentally, but that relaxation times can be estimated theoretically in terms of collisions between molecules and the properties of individual molecules. Conditions are different with regard to liquids: here the experimental evidence is abundant[1,2] but relaxation times cannot be derived from first principles, that is in terms of the properties of individual molecules and of the interactions between them.

No exact theory of the liquid state has yet been developed, even for the equilibrium properties of simple liquids. The problem has been approached in two ways, by treating a liquid as a compressed gas, or as a disordered solid, and both approaches have led to some successes. In empirical terms the liquid state may be described by a number of correlations between measured physical quantities. Two correlations between the properties of solids and liquids at comparable temperatures and pressures, not too close to the critical point, are simple and striking. Heats of vaporization are usually appreciably larger than heats of melting, which indicates that the cohesive energy of liquids is not much less than that of solids. Besides, the specific volume does not change greatly on melting, while it changes much more on evaporation, which shows that liquid molecules are in close touch. All this means that strong forces exist between molecules in liquids. However, at the same time the negligible shear strength of liquids indicates that molecules are not permanently constrained with regard to their position or orientation, and liquids have no long range order.

In the absence of an electrical field a dipole which is a feature of a molecule in a liquid may, averaged over a period of time, point in any direction of space with equal probability. This fact greatly simplifies the theoretical treatment of dipolar polarization, and liquids are the acknowledged testing ground for theories of the orientational polarization and its relaxation. However, it does not follow that all dielectric relaxation processes in liquids are best described in terms of the orientation of permanent dipoles.

A relaxation time characterizes the damping of some process. The processes in liquid which involve changes of dielectric polarization may be classified in various ways. In the present monograph we shall use a classification based on the experimental evidence of ultrasonic relaxation, originally suggested by Pinkerton and discussed fully by Herzfeld and Litovitz[3].

A. A CLASSIFICATION OF LIQUIDS

Herzfeld and Litovitz consider three groups of liquids according to the absorption coefficient α which characterizes the absorption of sound waves according to

$$A = A_0 \, e^{-\alpha x + i\omega \left(t - \frac{x}{c}\right)} \tag{11.1}$$

where A is the amplitude of the pressure or other relevant quantity, A_0 a constant, c the velocity of sound. The absorption coefficient α can be calculated from the viscosity and heat conduction of a liquid if one assumes that the only loss mechanism is as follows. A sound wave consists of crests and troughs of pressure; at the crests the temperature is increased, at the troughs decreased by adiabatic compression or expansion. Dissipation of free energy occurs when heat conduction irreversibly equalizes these temperature differences.

The ratio of the measured absorption coefficient α and the calculated classical value may be used to characterize three classes of liquids. Table I

TABLE I
Ultrasonic absorption of liquids

Liquid	Temperature °C	$\dfrac{\alpha}{\alpha_{\text{Classical}}}$	$\dfrac{1}{\alpha} \cdot \dfrac{\partial \alpha}{\partial T}$
		Group 1 (classical liquids)	
Hg	27	1·2	
A	−188	1·0	
N_2	−199	1·1	
		Group 2 (Kneser liquids)	
CS_2	25	1150	
CH_3I	2	24·7	0·0087
C_6H_6	22	100	0·011
C_6H_5Cl	25	15	0·0072
		Group 3 (associated liquids)	
H_2O	15	2·9	−0·036
CH_3OH	2	3·0	−0·011
$nC_5H_{11}OH$	28·6	1·8	−0·014
Glycerol	50	1·78	−0·05

(abbreviated from Herzfeld and Litovitz[3]) shows this ratio as well as the temperature coefficient of α. It may be seen that classical liquids (Group 1) exist, but are restricted to monatomic liquids and liquefied "permanent" gases. The vast majority of liquids are not classical, but show "excess" absorption. The non-classical liquids fall into two groups. Group 2 in the table are[3] called "Kneser" liquids and have large excess absorption. In general terms this can be attributed to a slow exchange of energy between internal and external degrees of freedom, although this interpretation is not universally accepted. The internal degrees of freedom are vibrations, oscillations, etc. taking place within individual molecules, while the external degrees of freedom are translations and rotations of the molecules as a whole, as has been mentioned for the case of gases. The magnitude of the ratio listed in the table is striking because it shows that even quite simple molecules are dynamically very different from hard spheres.

Group 3 differs from Group 2 by the sign of the temperature coefficient of α. The liquids in this group are "associated" in the sense that their molecules form short-lived associations with each other, so that the liquid may be considered as a mixture of different "polymers" the quantities of which are in dynamic equilibrium. The ultrasonic relaxation of such liquids may be interpreted by stating that the structure is pressure dependent, and that it adapts itself to a change of pressure with a time delay. The outstanding example of an associated liquid is water.

In practice, the division between liquids of Groups 2 and 3 is not sharp, in that an associated liquid may have slowly responding internal degrees of freedom, and associations may be more or less specific.

B. THE RELAXATION OF DIPOLE ORIENTATION IN LIQUIDS OF GROUP 2

The accumulated evidence on dielectric relaxation in polar non-associated liquids and in dilute solutions in such liquids as solvents may be interpreted in terms of the orientation of individual dipolar molecules. The simplest model for a dipolar molecule is a sphere which carries in its centre an infinitesimally small dipole. It is not obvious whether this should be considered as a model for a Group 1 or a Group 2 liquid. If the sphere is assumed rigid the model is classical but has no real counterpart. Real liquids with dipolar and nearly spherical molecules belong to Group 2, that is molecules are not rigid. This consideration may not be important, and models assuming rigid spheres or other shapes may represent dielectric properties correctly since for gases the slow relaxation of internal degrees of freedom does not seem to affect the order of magnitude of the dielectric relaxation time (see Table I). Nevertheless, the assumption of rigidity is in general not realistic.

Debye's original treatment of the relaxation of a dipolar liquid[4] (see Chapter 2) is in microscopic terms only up to the point where the friction constant has to be related to the properties of a given liquid. At that stage Debye assumes as a first approximation that the dipolar molecule behaves like a macroscopic sphere and obeys Stokes' law which refers to the sinking of a sphere in a liquid sufficiently viscous to prevent acceleration. This leads to a relaxation time

$$\tau = \frac{4\pi}{kT}\eta r^3 \qquad (11.2)$$

where r is the radius of the sphere and η the shear viscosity. The shear viscosity is measured by allowing a liquid to flow slowly through a tube, and by measuring the discharge as a function of pressure difference[5]. In those conditions the liquid at the inner surface of the tube may be taken to adhere to the surface, and friction takes place only within the liquid itself. Equation (11.2) does not imply any knowledge about the nature of the surface of the supposed macroscopic sphere, or of the way in which it fits into the liquid.

Debye's first-order approximation (11.2) should be realistic for very large dipolar molecules of approximately spherical shape. It can be checked in two ways, by comparing the radius calculated from the formula with the actual dimensions, and by examining whether the measured relaxation time is single or distributed. Table II shows a comparison of relaxation times calculated from Debye's equation with measured data for large dipolar molecules dissolved in non-polar solvents with relatively small molecules. The table is based on results by Meakins[6]; further data appear in Illinger's[2] review.

TABLE II

Solute	Solvent	Molecular volume $\times 10^{24}$ cm^3 of solute	of solvent	Viscosity (Centipoise)	$\tau_{exp} \times 10^{11}$	$\dfrac{\tau_{exp}}{\tau_{calc}}$
Cholestenone 0·005 M	Benzene	406	84	0·672	23·3	0·9
	decalin	406	154	2·71	67·7	1·2
	nujol	406		215	2730·0	2·4
Gammexane 0·15 M	Benzene	196	84	0·682	4·67	2·1
	decalin	196	154	2·67	7·93	5
Lupenone 0·1 M	Decalin	445	84	3·057	63·7	1·6
O-Dichlorobenzene 0·2 M	Decalin	116	154	2·565	1·53	14
	nujol	116	large	156	1340·0	520

Figure 1 illustrates the distribution of relaxation times for cholesterone solutions investigated by Meakins[6], as well as the structural formula of the solute. The molecule is not of spherical shape; Meakins replaces it in equation (11.2) by a sphere of equal volume. Figure 1 shows that the solutions in question do not have a single relaxation time—the experimental curves deviate from the Debye curves of best fit which are drawn in dotted lines. However, the deviation is not very large for the solutions in the less viscous solvents. Gammexane in benzene has according to Meakins a single relaxation time, within the limits of experimental error. Gammexane is the gamma isomer of $C_6H_6Cl_6$, that is a derivative of cyclohexane and more nearly spherical than cholestenone. Among other fairly large molecules in dilute solution giving single relaxation times are the symmetrical quaternary salt tetra-n-butyl ammonium picrate ((Bu_4N)picrate) in dioxane and benzophenone ($C_6H_5COC_6H_5$) in benzene investigated by Williams[7]. Pitt and Smyth[8], investigating two symmetrical very large molecules (molecular volumes 2400 and

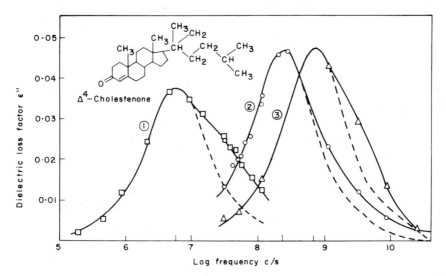

FIG. 1. Effect of varying solvent viscosity on dielectric properties of Δ^4 cholestenone of 0·05 M concentration at 20°C (from Meakins[6]). 1. Nujol; 2. Decalin; 3. Benzene.

2570×10^{-24} cm³ respectively), a metal free porphyrazine complex and its iron derivative, in benzene find a "fairly wide" distribution of relaxation times, while τ_{calc}/τ_{exp} = 1·35 and 0·58 respectively for these molecules of nearly ellipsoidal shape.

While the Debye equation (11.2) holds fairly well for large spherical molecules in solvents with appreciably smaller molecules, it does not hold at all well for the case where the solvent molecule is larger as may be seen for dichlorobenzene in Table II. The molecular radius calculated from the dielectric relaxation time according to the Debye equation is normally too small, i.e. the molecule responds faster than follows from the macroscopic viscosity and Stokes' law. Besides, the relaxation time is not a function of the viscosity of the solvent alone but depends both on solvent and solute.

There are two ways possible for proceeding beyond the first-order approximation used by Debye, one exact and the other approximate. In terms of the theory of transport processes dielectric relaxation times are coefficients among many others which may be calculated if the dynamic properties of a liquid are known in detail. An exact theory would give quantitative relationships between viscosities, relaxation times, conductivity for heat, etc. in terms of molecular properties, pressure and temperature. No such theory exists but Kirkwood[9] attempts it in terms of fluctuations. This theory is difficult and does not get beyond a general scheme for taking into account forces between molecules and a few general conclusions. One simple conclusion is that the frictional constant between solute and solvent is independent of the mass of the solute molecule if this is large compared with the solvent molecules.

For practical purposes an exact theory is as yet non-existent. This has led to many attempts to improve Debye's approximation while retaining a correlation between the dimensions of the solute molecule and the macroscopic viscosity of the solvent. Only a selection will be given here of the theoretical models that have been devised to this end. A more complete review is given by Illinger[2].

A simple improvement of the fit with experiment is given by taking into account the shape of the dipolar molecule by approximating it to a rigid ellipsoid rather than a rigid sphere while retaining the macroscopic viscosity. This model, which seems justifiable for large dipolar molecules, in a mobile solvent, was first proposed by Perrin[10] who found that it led to three distinct relaxation times. However, this result is valid only subject to certain somewhat artificial qualifications. Perrin considers Brownian motion and defines friction constants subject to the condition that during the time interval in question an ellipsoid changes its orientation with respect to a fixed coordinate system only slightly. In those circumstances, which might refer to a macroscopic ellipsoid subject to a force of short duration, three principal relaxation times may be defined according to the orientation of the three axes with respect to the field. Fischer and co-workers[11] tabulated numerical factors for dipolar ellipsoids from which equation (11.2) is modified to give three relaxation times

$$\tau_i = \frac{4\pi}{kT} \eta . abc . f_i \qquad (11.3)$$

where a, b and c are the axes of the ellipsoid and f_i is a function of the axial ratios.

Perrin's[10] and Fischer's[11] treatment was devised for use with experiments at frequencies where $\omega\tau \ll 1$. This condition is inconsistent with the assumption made about the short duration of the force acting on the ellipsoid, but in practice formula (11.3) has been found useful. Fischer[11] and co-workers assumed three Debye peaks with the three relaxation times τ_i, of magnitude dependent on the components of the dipole moment along the three axes, and interpreted the increase of $\varepsilon''(\omega)$ at low frequencies as due to the superposition of these peaks.

While the treatment in question is justifiable in its experimental context, the derivation of formula (11.3) with its three relaxation times is untenable for frequencies where $\omega\tau$ is near unity. The dielectric relaxation time is the regression time of the fluctuations of dipole orientation. When the duration of a cycle is of the order of the average τ, molecules cannot be meaningfully subdivided into groups with different orientation, where the different orientation is assumed to hold over periods longer than τ. Separate relaxation times for ellipsoids of different orientation are not characteristic of separate processes. Equation (11.3) may be used to estimate relaxation times by averaging over the three values, as done by Fischer[11], but it is not a valid derivation of a distribution of relaxation times in terms of fluctuations.

Another model, by Wirtz[12] and co-workers, also does not consider thermal motion on a microscopic scale. Wirtz retains a rigid spherical dipolar molecule and modifies equation (11.2) on the grounds that Stokes' law ceases to be realistic for a sphere whose radius is not large compared with the size of surrounding molecules. Wirtz assumes that the solvent molecules form discrete shells of radius $r_2 + nr_1$ about the centre of the dipolar sphere where r_2 is the radius of the solute, r_1 that of the solvent molecule and n an integer. Friction then takes place between shells, and the macroscopic coefficient of friction η in equation (11.2) is modified by multiplication with a function of $\frac{r_1}{r_2} = x$, namely

$$\frac{\eta_{\text{micro}}}{\eta} = f(x) = \frac{1}{6x + (1 + x)^{-3}}. \tag{11.4}$$

Equation (11.4) is an improvement on the Debye equation for several solvent-solute combinations[2].

Improved fit with measured data can be obtained also by some frankly empirical formulae, some with a disponible parameter. Le Fèvre and collaborators give two empirical formulae without disponible parameters which give good fit for several solvent-solute combinations and a fair representation of the temperature dependence of the relaxation times. One of these formulae, by Le Fèvre and Sullivan[13]

$$\tau = \eta \frac{\pi}{2kT} ABC\, e^\Delta .(\varepsilon + 2)^{-1} \tag{11.5}$$

involves the ellipsoid with axes A, B and C defined by the polarizability of the solute, Δ is the depolarization factor and ε the dielectric constant of the solvent. The introduction of these parameters implies a connection with the resistance to deformation of the molecules and with electrostatic interaction within the solvent.

A more fundamental approach by Hill (1954) takes into account the fact that both viscous flow and dielectric relaxation are consequences of thermal motion. The argument is based on a dynamic model by Andrade[15] which assumes that a molecule in a liquid spends most of its time vibrating about some temporary equilibrium position which varies only slowly with time. Occasionally a more energetic process takes place which is analogous to a collision in a gas, and the transfer of momentum which is responsible for friction is largely due to this process. Theories of liquid viscosity based on models of this kind are reviewed by Herzfeld and Litovitz[3], and are found to be successful in explaining experimental data.

The simplest model of the type discussed is probably the bistable model, treated in Chapters 2–4. This has the advantage of leading to a simple dependence of the relaxation time on temperature, the dependence characteristic for a rate process (see equation 4.35). The viscosity of Class II liquids is normally exponentially dependent on reciprocal temperature in the way of a rate process.

Hill[14] considers the viscosity of a mixture of two liquids A and B and finds an expression for the "mutual" viscosity of the two liquids

$$\eta_{AB} = \tfrac{1}{3} C_{AB} \frac{v_A + v_B}{\sigma_{AB}} \cdot \frac{m_A m_B}{m_A + m_B} \tag{11.6}$$

where m_A and m_B are the masses of the two kinds of molecules and v_A and v_B are the characteristic frequencies for their vibrations about temporary equilibrium positions. The factor C_{AB} is a constant characteristic for momentum transfer between A and B molecules while σ_{AB} is the mean distance between the two kinds of molecule. The average viscosity of a liquid mixture is derived in terms of η_{AB} and the viscosities η_A and η_B of the two liquids, and is found to give good agreement with experiment for various mixtures.

The dielectric relaxation time of a polar molecule B dissolved in a non-polar liquid A may be derived from the mutual viscosity η_{AB}, provided that the process causing dielectric relaxation is the same process which causes viscous flow. Hill shows that with the assumptions made

$$\tau = \frac{3}{kT} \cdot \frac{I_{AB} I_B}{I_{AB} + I_B} \cdot \frac{m_A + m_B}{m_A m_B} \cdot \eta_{AB} \sigma_{AB} \tag{11.7}$$

where I_B is the moment of inertia of the polar molecule about its centre of mass and I_{AB} is the moment of inertia of the molecule A about the centre of mass of B at the instant of a "collision". The molecules are taken to interact as rigid bodies throughout, i.e. momentum is not transferred to internal degrees of freedom.

Table III shows a comparison between the fit with experiment of the equations discussed so far for a few dilute solutions of compact spherical or ellipsoid molecules at room temperature. The dilution of the solutions implies that dipolar interaction may be neglected, while the molecular shape and the character of the solvent implies that the main process of relaxation is the orientation of dipoles. Further comparative tables are given by Illinger[2].

TABLE III
Relaxation times of dilute solutions of polar molecules

Solute	Solvent	observed	Relaxation time × 10^{11} sec calculated according to					Cole–Cole α	Reference
			equation (11.2)	equation (11.3)	equation (11.4)	equation (11.5)	equation (11.7)		
Anthrone	Benzene	2·48	15·3	10·0	2·31	2·84	8·8	0·14	16
Fluorenone	Benzene	1·99	13·2	9·08	1·95	2·46	8·3	0·13	16
Phenanthrene-quinone	Benzene	3·13	14·6	10·7	2·28	2·72	9·5	0·15	16
Ferric octaphenyl porphyrazine chloride	Benzene	70·6	41	88·9	22·3	20·0	264·0	"rather large"	8,16
O-Dichlorobenzene	Decalin	1·53	22·1				4·4		5
Phenanthrene-quinone	Dioxane	9·18					24·6		16

Table III and the more ample selection of data in references 1, 2, 6 and 16 indicate that the Hill equation (11.7) gives better agreement with experiment than the Debye equation (11.2) (interpreted by replacing the molecule by a sphere of equal volume) for small molecules, but worse agreement for very large molecules. Equations (11.4) and (11.5) give very good agreement with experiment in the first three solvent solute combinations of the table. However, the individual formulae fit some solvent–solute combinations better than others. In the context, it seems interesting to note that the ultrasonic relaxation time of benzene at 20°C is $\tau_u = 28·8 \times 10^{-11}$ sec.

All the formulae quoted above correlate the dielectric relaxation time with a viscosity. Hill[17] states that this correlation does not necessarily hold. The process relevant for friction is an energetic interaction, a kind of collision, between liquid molecules in which they become temporarily locked with one another. Such an association is permanent in a solid with the result that internal friction in a solid is exceedingly high compared with liquid friction. However, dielectric relaxation is in some cases possible in a solid, as may be seen in Meakins' results[6] for liquid and solid lupenone. Figure 2 shows that the freezing of this substance frustrates molecular rotation, but does not prevent some kind of motion of a polar attachment of the large molecule. There is no reason for assuming that the relaxation time of the latter process in the liquid should be closely linked with the viscosity. The rotation of com-

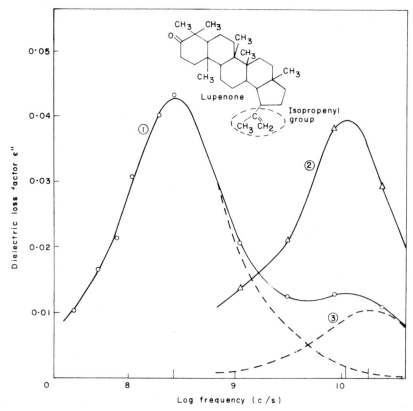

FIG. 2. Dielectric absorption of lupenone at 20°C in decalin solution and in the crystalline state (from Meakins[6]). 1. 0·1 M lupenone in decalin molecular rotation; 2. Solid lupenone (isopropenyl group); 3. Isopropenyl group.

plete spherical molecules may also be possible in a solid, as will be discussed in Chapter 12. In general there is no certainty that the viscosity, the coefficient of diffusion and the dielectric relaxation time in a liquid all refer to the same process, although the Debye approximation[4] which links these three coefficients (see Chapter 2) may be expected to be roughly valid in a mobile liquid.

The discussion so far was concerned with dilute solutions of polar molecules in non-polar solvents, where the relaxation time may theoretically depend on the dielectric constant of the solvent but not significantly on dipolar interaction between the solute molecules. According to the theoretical arguments sketched in Chapter 2 and discussed more fully in Chapter 8, the dielectric relaxation time should be lengthened by dipolar interaction. Miller

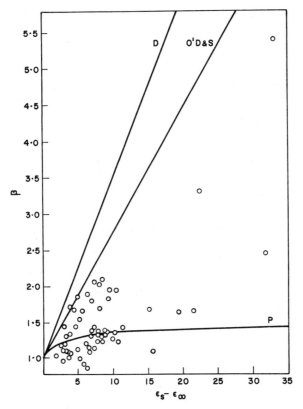

FIG. 3. The enhancement of the dielectric relaxation time of compounds, β_1 compared with $\varepsilon_s - \varepsilon_\infty$ (from Miller and Smyth[18]). Circles represent experimental values. Curve D represents equation (2.65), P equation (8.29) and $O'D.$ & $S.$ a formula by O'Dwyer and Sack[19].

and Smyth[18] explore the relationship between intrinsic and extrinsic relaxation times by the following method: they select pairs of liquids of very similar structure, one with and one without dipole moment. They assume that in the absence of the dipole, dielectric relaxation times in these liquids as solvents would be related as their reciprocal viscosities. The enhancement β due to the inner field in the concentrated dipolar liquid is calculated as

$$\beta = \frac{\tau_2 \eta_1}{\tau_1 \eta_2} \qquad (11.8)$$

where 2 refers to the polar and 1 to the non-polar liquid. For spherical molecules the two viscosities are assumed equal. For the simplest case, originally envisaged by Debye, β would be given by equation (2.65) which assumes that the dipolar liquid responds like a continuum. The correct equation for the static dielectric constant of a dipolar liquid is the Onsager equation (8.20) or Fröhlich's more general expression (8.24). The enhancement β of the relaxation time according to the dynamic generalization of the Onsager equation by Powles and by Glarum would be given by equation (8.29), while Scaife[20] disagrees with this result as outlined in Chapter 8.

Miller and Smyth tabulate β according to equation (11.8) for a number of pairs of liquids as a function of $\varepsilon_s - \varepsilon_\infty$ and find Fig. 3 as conspectus of their data. The graph contains a theoretical expression according to O'Dwyer and Sack[19] (*O'D & S*), which gives results nearer to the Debye curve (*D*), than to the Powles curve (*P*). The point with $\tau_2/\tau_1 = 5\cdot 4$ corresponds to the pair of liquids CBr_2Cl_2 and $(CH_3)_2C(NO_2)_2$, the first one being non-polar. If this value is corrected for viscosities (a procedure rejected by the authors of the graph for spherical molecules), the value of β comes down to 4·3, but is still high above the Powles curve. The next highest point at $\beta = 3\cdot 31$ is corrected for viscosity and corresponds to toluene versus benzonitrile.

The wide spread of points in Fig. 3 confirms Scaife's conclusion that no law of general validity can be found for the relationship between intrinsic and measured relaxation time in liquids. The Powles relationship seems to be nearer to the average of the experimental data than are the other two formulae.

C. ASSOCIATED LIQUIDS

Associated liquids differ from other liquids in that the forces between their molecules are stronger and/or more directional. They have in common the ultrasonic characteristics[2] summarized in Table I; that is they have an excess absorption α/α_{class} which is between unity and three and is not much dependent on temperature. The negative temperature coefficient of α is due to thermal expansion since the data refer to constant pressure. Another characteristic of associated liquids is the behaviour of the viscosity, in that $\log \eta$ is not related linearly to $1/T$ as in Class II liquids and as is to be expected for a single rate process.

Dielectric relaxation in an associated liquid can be treated as a process of dipole orientation, in that the dipoles attached to molecules fluctuate in their orientation. However, an individual dipole is now part of a cluster of molecules even though it changes its partners frequently. In an associated liquid the formation and dissolution of clusters is an important process and dipole orientation cannot be easily separated from it.

The most important associated liquid is water. Water is also unique in that it has a density maximum at 4°C; besides, water has many other unusual properties such as a very high specific heat (unity at 15°C) and a very large dielectric constant ($\varepsilon = 78.2$ at 25°C). The dielectric properties of water have been reviewed by Hasted[21].

The molecule H_2O has a triangular shape determined by the angle between the two oxygen orbitals which bond to hydrogen (see Fig. 4). Apart from the

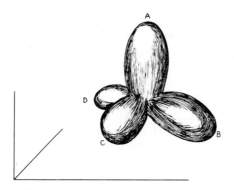

FIG. 4. Schematic representation of bonding and lone-pair electron clouds in an isolated H_2O molecule. A and B represent electron pairs in O—H bonds where one of the electrons is contributed by O and the other by H, the proton being embedded in the charge cloud. C and D represent lone-pair charge clouds. Note that A, B, C and D make almost tetrahedral angles with each other.

two bonding orbitals, oxygen in water has also two "lone pairs" of electrons which project, as shown in the figure, so that the six outer electrons of the oxygen atom form four projections, at angles of approximately 109° to each other. In ice each oxygen atom is surrounded tetrahedrally by four other oxygens, a configuration which comes close to satisfying the bond angles of the free molecule (see Fig. 5). Hydrogen atoms are situated along the lines connecting oxygens and form hydrogen bonds between the oxygen atoms to which they are bonded by a covalent bond and the neighbouring oxygen atom which attracts them by virtue of a projecting lone pair of electrons. A hydrogen bond signifies that a hydrogen atom (proton) joins two electro-negative atoms (for example oxygen) together by a combination of dipolar and quantum mechanical forces[22,23,24]. The energy needed to break a hydrogen bond is of the order 5 kcal/mole (0.2 eV) and thus much less than for a typical covalent bond which has an energy of the order 4 eV.

In ice, the dielectric properties of which will be discussed in Chapter 15, all or almost all possible hydrogen bonds are formed. In consequence ice has a large molecular volume since the tetrahedral arrangement is wasteful of space. Water has a smaller molecular volume than ice because a considerable fraction of its possible hydrogen bonds are "broken" which may mean that they are strongly bent but still have some bonding effect. The fraction of broken bonds increases with rising temperature, and this explains the density maximum.

The kinetic process responsible for dielectric relaxation may be pictured as follows. Water consists of fluctuating clusters of bonded molecules, with non-bonded molecules between them. Individual molecules frequently pass from one cluster to another, and they are more or less free to change their orientation according to the number of hydrogen bonds they form[21]. A singly bonded molecule can rotate about a hydrogen bond without energy of activation, while a molecule forming three or four bonds is fixed by them, unless it obtains an energy sufficient to break two or three bonds.

TABLE IV

Temperature (°C)	$\tau_D(H_2O)$ sec × 10^{12}	$\tau_D(D_2O)$ sec × 10^{12}	$\tau_u(H_2O)$ sec × 10^{12}
0	17·8		
10	12·7	16·6	
20	9·6	12·3	2·1
40	5·95	7·6	
60	4·05	4·92	

The fluctuation of the polarization in water, and therefore the static dielectric constant, can be calculated on the basis of a detailed assessment of the surroundings of a dipole. The Onsager equation (8.20) on the basis of $\varepsilon_\infty = n^2$, where n^2 is the squared refractive index, gives only $\varepsilon_s = 31$ for 25°C as compared with the measured $\varepsilon_s = 78·3$, but Oster and Kirkwood[25] were able to deduce a value close to the measured one by a treatment which may be considered as a special case of the general procedure indicated by equation (8.24). This calculation implies cooperative ordering, i.e. it implies that neighbouring dipoles are to some extent parallel to one another. However, according to Hill[26], the magnitude of ε_s can be explained on the basis of the Onsager equation if $\varepsilon_\infty = 5$, a value which is acceptable on the basis of recent evidence.

The dielectric relaxation of water has been measured over a representative frequency range[21,27] (although ε_∞ is still beset with some uncertainty). Table IV gives the dielectric relaxation time of H_2O and D_2O as a function of temperature, as well as a figure for the ultrasonic τ_u.

The activation energy of the relaxation time of water[2] is 4·5 kcal/mole and identical with the activation energy of self-diffusion and of the viscosity. This suggests that the fundamental process is the same in all three cases.

The difference between the ultrasonic and dielectric relaxation times might be attributed to electrostatic interaction. If so, then τ_u is a more accurate measure of the life time of clusters than is τ_D.

The dielectric relaxation time of water has a very narrow distribution with a Cole–Cole parameter[27] of only $\alpha = 0.02 \pm 0.007$. This is interesting because it implies that clustering is a practically random process, and that it is not possible to divide water molecules into groups which remain different over periods longer than the average τ_D. One might expect that the probability for the reorientation of a water molecule would be reduced while it was held inside an ice-like cluster.

The almost single relaxation time of water may be connected with the nature of its infra-red vibrations. The OH stretching vibration whose fundamental frequency in water is around $v = 1 \cdot 10^{13}$ c/s is strongly perturbed by hydrogen bonding[24], which means that hydrogen performs excursions of anomalously large amplitude and low frequency. The intense OH stretching vibration and related vibrations may keep the clusters on the move in a random way. Table IV contains evidence to the effect that thermal vibrations are important. The ratio of the normal stretching vibrations of OH and OD is $\sqrt{2}$, and the ratio of the relaxation times of D_2O and H_2O approaches this value with decreasing temperature. Hill[26] discusses the infra-red vibrations of H_2O and D_2O in connection with the large value of ε_∞, which is, of course, a sign of intense infra-red absorption. Even a value of only $\varepsilon_\infty = 4$ is much larger than the square of the refractive index which is $n^2 = 1.79$.

The magnitude of τ_D is in good agreement with equation (11.2). However, agreement with the Debye formula need mean no more than that the viscosity and τ_D are both due to the same process.

There are in principle two processes present in the relaxation of water, one the fluctuation of the polarization, and the other the dissociation of water into ions. The second of these processes is of a type which has been investigated by ultrasonics, for the case of electrolytes dissolved in water. Figure 5 shows measured data for the ultrasonic absorption in a solution of $MgSO_4$ in water at 20°C, from Eigen et al.[28].

The figure shows two separate practically pure Debye peaks, corresponding to two different processes. Eigen attributes these processes to reactions which may be schematically written

$$Mg^{++} + OH^- = MgOH^+ \quad \text{(a)}$$

$$SO_4^{--} + H^+ = HSO_4^- \quad \text{(b)}$$

The first process is the slower one. A third process which must occur simultaneously

$$OH^- + H^+ = H_2O \qquad (c)$$

is still more rapid and does not appear on the figure. The speed of the neutralization reaction c is high in this case because the concentration of H^+ and OH^- ions is kept high by the presence of the electrolyte. Eigen and de

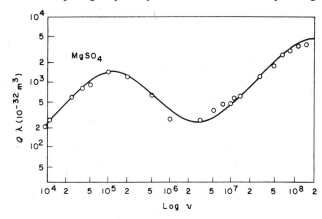

FIG. 5. Ultrasonic absorption for $MgSO_4$ in aqueous solution. The product of absorption cross section Q and wavelength λ is plotted against the logarithm of the frequency (from Eigen et al.[28]).

Maeyer[29] estimate that for purest water the relaxation time of the neutralization reaction is $2\cdot6 \times 10^{-5}$ sec at 25°C, that is, this process is much slower than the dipole relaxation. It might in principle lead to a distribution of relaxation times if clusters near ions behave differently from clusters far from ions.

The dielectric relaxation time of water is influenced by the addition of electrolytes as well as non-electrolytes. The significance of this influence has been interpreted with regard to the structure of clusters[21].

While the dynamic properties of water can be understood at least in general terms on the basis of the properties of molecules and the interactions between them, other associated liquids have not been studied in as fundamental a way. However, a great deal of information is available, particularly with regard to alcohols. The series CH_3OH, C_2H_5OH, etc. shows a gradation from water-like behaviour of the lowest members of the series to the behaviour of a long-chain paraffin.

Alcohols have more or less wide distributions of relaxation times. In many cases their Cole–Cole plot is of the "skewed arc" type (equation 7.21) although this equation only characterizes the general shape. Arc plots tend to resemble the last diagram of Fig. 2 of Chapter 7 which is obtained by the superposition of a slow and a fast relaxation process. The arc plot for[30] n-heptanol is characterized by a slow process with an almost pure Debye semi-circle, and some faster processes. The latter becomes more important as the length of the hydrocarbon chain increases. Brot[30] attributes the slow process to cooperative clustering of OH groups, which in the case of normal alcohols involves the formation of linear chains of dipoles (see Chapter 15). He attributes the fast processes to OH groups which are in anomalous positions, i.e. not associated, or terminal links of a chain. This implies that the anomalous character of the anomalous dipoles persists for a time which is long compared with the relaxation times of the fast processes.

The generalized shape of a relaxation spectrum in the skewed arc plot (see Fig. 5 of Chapter 7 and equation 7.24) may be interpreted according to Glarum[31] by a model which implies that the relaxation process is not random. Glarum assumes that the re-orientation of a molecule is more likely immediately after one of its neighbours has relaxed than it is at an arbitrary time. He expresses this by writing for the decay function of dipole orientation

$$\psi(t) = e^{-\alpha t}.(1 - p(t)) \qquad (11.9)$$

where α is a constant characterizing the decay for a dipole inside an ordered cluster and $p(t)$ is the probability that a defect reaches an arbitrary dipole at a time t and acts in such a way as to bring about its immediate relaxation. This model presupposes that a decay function $\exp(-\alpha t)$ can be meaningfully defined for the inside of clusters and that defects can be treated separately. Glarum's model leads to the equation for a skewed arc on the basis of specific assumptions about $p(t)$.

The skewed arc plot is interesting with regard to the definition of association in liquids. Not only hydroxyl compounds but also nitrides (for example HCN) and other compounds containing hydrogen as well as an electronegative atom may form hydrogen bonds and are thus liable to associate in the liquid state. However, the skewed arc plot occurs also in supercooled liquids which are not markedly associated in the temperature range where the liquid state is stable. Denney[32] finds a skewed arc plot for several supercooled, branched, hydrocarbon halides, for instance Isobutyl bromide, although according to Smyth[33] this compound has a single relaxation time in the normal liquid range. The Glarum[31] model was devised for these cases and it seems apt for them, since a supercooled liquid is likely to contain ordered regions which persist ordered for considerable periods.

D. LIQUIDS OF VERY HIGH VISCOSITY

Certain associated liquids supercool easily and reach in a supercooled condition viscosities which class them as glassy solids rather than normal liquids. The most interesting liquids in this group are glycerol, $C_3H_5(OH)_3$

and related liquids, and mixtures of chlorinated diphenyls. The latter are of practical interest, and known under trade names such as "aroclor" or "pyranol".

Glycerol is very strongly hydrogen bonded. It is unusual in that the ultrasonic and dielectric relaxation times are equal[3] indicating either that electrostatic interaction does not enhance the relaxation time or that the intrinsic dielectric relaxation time is faster than the ultrasonic one. A comparison of τ_u and τ_D for several associated liquids by Herzfeld and Litovitz[3] (p. 501) indicates that the relationship between these two quantities is not straightforward.

The relaxation time of glycerol is well described by a skewed arc plot[34]. At a temperature of $-50°C$ which is far below the melting point of $17.9°C$, the parameters are[34] $\tau_0 = 1.25 \times 10^{-4}$ sec and $\beta = 0.603$. This is a wide distribution, as might be expected; the ultrasonic[3] distribution is not as wide as the dielectric one. The relaxation time depends strongly on temperature, but not in the way of a normal rate process. (Incidentally τ_0 in the Cole–Davidson equation (7.24) for a skewed arc is not the average relaxation time τ_D, though closely related to it; for $\beta = 0.6$, $\tau_0 = 1.5\tau_D$.) It has been found that

$$\tau_0 = A\, e^{\frac{B}{T-T_\infty}} \qquad (11.10)$$

for glycerol and related liquids, where A, B and T_∞ are constants. The viscosity for glycerol follows the same law with the same B and $T_\infty = 132°K$ for both. In consequence equation (11.2) is satisfied, over a wide range of temperatures, but[34] with a value for r which is too small by an order of magnitude. Equation 11.10 suggests that the ordering in the liquid increases with falling temperature, until relaxation becomes infinitely slow at T_∞. In practice[34] the viscosity rises to the order 10^{13} poises some $20°$ to $40°$ above T_∞, so that the liquid becomes a glass.

The viscosity and relaxation time of glycerol is strongly affected by small amounts of water. Measurements[35] indicate a linear dependence of log τ on the concentration up to 47% of water and suggest that there should be an approximately linear dependence up to 100% of water. Litovitz[35] and co-workers ascribe these results to the formation of composite clusters of water and glycerol. In a series of papers[36a,b,c] Litovitz and co-workers relate the average dielectric relaxation time $\bar{\tau}_D$ of mixtures of two associated liquids to the average relaxation time $\bar{\tau}_s$ of their shear viscosity. They find[36c] for mixtures of glycerol and n-propanol that $\bar{\tau}_D/\bar{\tau}_s$ varies non-linearly with concentration, rising from 2 for glycerol to 80 for n-propanol. The width of the distribution narrows as the ratio $\tau_D/\bar{\tau}_s$ becomes larger. The authors suggest an interpretation of their data in terms of different local free volumes relevant for dielectric and ultrasonic relaxation. For mixtures of an associated and a non-associated liquid two resolved relaxation peaks are observed in many cases[37].

Associated liquids may be considered to consist of ordered and disordered regions and since the latter are likely to be of lower density the balance between order and disorder is likely to be pressure sensitive in the sense that order increases with pressure. Indeed the relaxation time increases with pressure for glycerol, as well as for the aroclors. The pressure dependence data are summarized for both types of liquid by Perls and Wilner[38], in conjunction with work on certain aroclors. For glycerol the change of τ_D brought about by altering the pressure is simply related to the change brought about by altering the temperature. The pressure needed to compensate a rise of $1\,°C$ is 200–250 kg/cm^2 over the whole temperature range. This ratio is for room temperature similar to the ratio of thermal expansion and compressibility. It would seem tempting to assume that τ_D may be a function of the volume only, i.e. that for constant volume τ_D may be independent of temperature. However, conditions are in general more complex. Gilchrist et al.[39] examined the pressure and temperature dependence of the molar volume of glycerol and propanol, and correlated them with the dielectric data. Measurements at constant volume are impracticable over a wide temperature range, but the authors estimate dielectric data for constant volume. They find that even for constant volume $\tau(T)$ does not behave according to a single rate process.

The chlorinated diphenyls are useful impregnants for paper capacitors because they combine a fairly high dielectric constant with chemical stability[40]. They are mixtures based on diphenyl, $C_6H_5 \cdot C_6H_5$ and have substituted chlorine atoms in varying numbers and positions. Under the trade name aroclor they are graded so that the last two figures of a four-figure code give the weight percentage of chlorine. For instance, type 1260 contains about six chlorine atoms, type 1242 about three.

The aroclors have moderately wide relaxation peaks of the skewed arc type[35]. The viscosity as a function of temperature does not obey equation (11.10) but shows a linear relationship between $\log \eta$ and $1/T^3$ at low temperatures. The ratio η/τ_D changes appreciably with temperature. The pressure dependence of the relaxation time is exceptionally large.

The complicated relaxation behaviour of the aroclors may be appreciated in terms of structure, insofar as one may expect that different forms of clustering may be possible and thermodynamically about equally favourable. The chlorinated diphenyls can twist about the carbon–carbon bond joining the two rings, and the average configuration of the two rings will in general be different for different isomers. The C—Cl groups are dipolar, and may form hydrogen bonds[24] with adjacent C—H groups. Since the aroclors are mixtures of isomers and of compounds with different chlorine constants, many patterns of short range order seem possible, and external variables such as the pressure may exert a strong influence on the structure.

REFERENCES

1. C. P. Smyth (1955). "Dielectric Behaviour and Structure." McGraw-Hill, New York.
2. K. H. Illinger (1962). "Dispersion and Absorption of Microwaves in Gases and Liquids." Progress in Dielectrics, Vol. 4. Academic Press, New York and London.
3. K. F. Herzfeld and T. A. Litovitz (1959). "Absorption and Dispersion of Ultrasonic Waves." Academic Press, New York and London.
4. P. Debye (1929). "Polar Molecules." Chemical Catalog Co. 1929, Dover Publications.
5. P. P. Ewald, Th. Pöschl and L. Prandtl (1936). "The Physics of Solids and Fluids." Blackie & Sons, London.
6. R. J. Meakins (1958). *Trans. Faraday Soc.*, **54**, 1160. Also (1958). *Proc. phys. Soc. Lond.*, **72**.
7. G. Williams (1959). *J. Phys. Chem.*, **63**, 534, 537.
8. D. A. Pitt and C. P. Smyth (1959). *J. Phys. Chem.*, **63**, 582.
9. J. G. Kirkwood (1946). *J. Chem. Phys.*, **14**, 180.
10. F. J. Perrin (1939). *J. Phys. Radium*, **5**, 497.
11. E. Fischer (1939). *Phys. Z.*, **40**, 645.
12. A. Gierer, A. Spernol and K. Wirtz (1953). *Z. Naturforsch.*, **8**a, 522, 532.
13. R. J. W. Le Fèvre and E. P. A. Sullivan (1954). *J. Chem. Soc.*, **2**, 873.
14. N. Hill (1954). *Proc. Phys. Soc. Lond.*, **67**B, 149.
15. E. N. da C. Andrade (1934). *Phil. Mag.*, **17**, 497.
16. D. A. Pitt and C. P. Smyth (1958). *J. Amer. chem. Soc.*, **80**, 1061. Also (1959). *J. Am. chem. Soc.*, **81**, 783.
17. N. Hill (1957). *Proc. Roy. Soc.*, **A240**, 101.
18. R. C. Miller and C. P. Smyth (1957). *J. Am. chem. Soc.*, **79**, 3310.
19. J. J. O'Dwyer and R. A. Sack (1952). *Austr. J. sci. Res.*, **A5**, 647.
20. B. H. P. Scaife (1965). E.R.A. Report No. 5032/1017.
21. J. B. Hasted (1961). "Progress in Dielectrics", Vol. 3, p. 101. Academic Press, New York and London.
22. L. Pauling (1960). "The Nature of the Chemical Bond." Cornell University Press, Ithaca.
23. C. A. Coulson (1957). *Research*, **10**, 149.
24. G. C. Pinnentel and A. L. McClellan (1960). "The Hydrogen Bond." W. H. Freeman, London.
25. G. Oster and J. G. Kirkwood (1943). *J. Chem. Soc.*, **11**, 175.
26. N. Hill (1963). *Trans. Faraday Soc.*, **59**, 344.
27. E. H. Grant, T. J. Buchanan and H. F. Cook (1957). *J. Chem. Phys.*, **26**, 156.
28. M. Eigen, G. Kurtze and K. Tamm (1953). *Z. Elektrochem.*, **57**, 103.
29. M. Eigen and de Maeyer (1955). *Z. Elektrochem.*, **59**, 986.
30. C. Brot (1957). Thèse, Masson et Cie, Paris.

31. S. H. Glarum (1960). *J. Chem. Phys.*, **33**, 639.
32. D. J. Denney (1957). *J. Chem. Phys.*, **27**, 259.
33. C. P. Smyth et al. (1948). *J. Am. chem. Soc.*, **70**, 4093. Also (1952). *J. Chem. Phys.*, **20**, 1121.
34. D. W. Davidson and R. H. Cole (1951). *J. Chem. Phys.*, **19**, 1484.
35. G. E. McDuffie, R. G. Quinn and T. A. Litovitz (1962). *J. Chem. Phys.*, **37**, 239.
36a. G. E. McDuffie and T. A. Litovitz (1962). *J. Chem. Phys.*, **37**, 1699.
36b. T. A. Litovitz and G. E. McDuffie (1963). *J. Chem. Phys.*, **39**, 729.
36c. R. Kono, T. A. Litovitz and G. E. McDuffie (1966). *J. Chem. Phys.*, **45**, 1790.
37. D. Denney (1959). *J. Chem. Phys.*, **30**, 1019.
38. T. A. Perls and L. B. Wilner (1960). *J. Chem. Phys.*, **33**, 753.
39. A. Gilchrist, J. E. Earley and R. H. Cole (1957). *J. Chem. Phys.*, **26**, 196.
40. J. B. Birks (ed.) (1960). "Modern Dielectric Materials." Heywood, London.

CHAPTER 12

Dielectric Relaxation in Crystalline Solids

The present chapter is concerned with various mechanisms whereby dipoles may re-orient in perfect or almost perfect crystals. It is not concerned with mechanisms where ionic or electronic defects are able to migrate over larger than molecular distances, i.e. mechanisms which imply the possibility of conduction phenomena. The chapter also excludes migrating flaws in hydrogen bonding, which will be discussed in Chapter 15.

In a perfect crystal each atom, ion, or molecule, has its place where it is held in position by the force fields of its neighbours. In perfect molecular crystals dipolar re-orientation is nevertheless possible if a dipolar molecule behaves as if it were a smooth sphere or cylinder. Behaviour of this kind is sometimes called "molecular rotation", although the rotation is of course not continuous. In ionic crystals limited movements of ions or other charge carriers are possible in the presence of flaws which imply only a small departure from perfection with regard to other physical properties.

A. DIPOLAR "ROTATION" IN MOLECULAR CRYSTALS

Globular Molecules

Quite a large number of organic molecules are "globular", i.e. approximately spherical in shape, and form "plastic" crystals, where molecules change their orientations freely in the solid state. The properties of such crystals have been recently reviewed[1]. Figure 1 shows a comparison of the

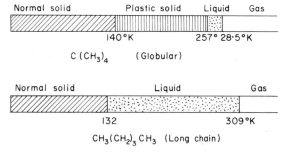

FIG. 1. Phases in equilibrium over a temperature range for two pentanes of overall formula C_5H_{12} (from Timmermans[1]).

phases formed by two compounds of very similar chemical composition, one globular and one not. The globular compound $C(CH_3)_4$ forms a plastic phase at a temperature close to the melting point of the non-globular one, and the liquid phase of the globular compound has a very short temperature range. The heat and entropy of melting in plastic phases is small, indicating a similarity between the thermal motions in plastic crystals and liquids. The low temperature phases formed by globular molecules are not plastic.

The crystal structures[2], infra-red properties[3] and dielectric properties[4] of globular molecules, in particular substituted methanes[5], have been widely investigated. The structures of the substituted methanes are usually cubic[2], and the infra-red spectra of plastic phases are more similar to those of the liquid than are the spectra of non-plastic phases[5]. However, the interpretation of these spectra is hampered by the lack of an adequate theory of the coupling between vibrational modes of adjacent molecules[3].

The presence of dipoles reduces the stable range of the plastic phases. While $C(CH_3)_4$ changes from a non-rotational to a rotational phase at 140°K, $(CH_3)_3CNO_2$ does so at 258°C and $(CH_3)_3CCN$ melts at 291°K without forming a plastic phase[5]. Clemett and Davies[5] correlate these differences in the transition temperature with the cohesive energies, to which the electrostatic interaction of dipoles makes a larger contribution in the non-rotational than in the rotational phase. However, although these authors also calculate the dispersion (van der Waals or London) forces, the transition temperatures cannot be calculated exactly.

Clemett and Davies[5] examine the dielectric properties of five substituted methanes in great detail. They find that the liquid phases obey the Onsager equation (8.21) quite well. On solidifying to a plastic phase the dielectric constant changes only slightly, but the temperature dependence of ε_s in the solid may differ from that in the liquid.

Clemett and Davies determine the dielectric relaxation times of solid rotator phases and corresponding liquids as well as the activation energies of the relaxation times. Table I gives a comparison of these data for four compounds, the relaxation times being extrapolated to the melting point.

The table shows that the relaxation times for the rotator phases are very

TABLE I

Compound	W(liquid)	W(solid)	τ (liquid)	τ (solid)
	in kcal/mole		$\times 10^{12}$ sec	
$(CH_3)_2CCl_2$	$1\cdot2_5$	1·4	12·7	13·6
$CCl_3 \cdot CH_3$	$1\cdot1_2$	(1·1)	11·0	12·4
$(CH_3)_3CNO_2$	$0\cdot8_1$	0·5	7·9	10·2
$(CH_3)_2CClNO_2$	$1\cdot4_6$	1·31	20·7	21·5

close to those of the liquids, and the activation energies are similar, which indicates that the motions causing relaxation are much the same. Clemett and Davies discuss this result with regard to the relevance of liquid viscosity (see Chapter 11). They also correlate the data with the specific volume of liquids and solids, by assuming a relationship analogous to one used in the theory of viscosity[6]

$$\frac{A}{\tau_0 T} = v - b \tag{12.1}$$

where τ_0 is the intrinsic relaxation time, v the molecular volume, A a constant and b the molecular volume of the "unexpanded" solid. The data can be fitted by straight lines according to equation (12.1), but the value of b comes out appreciably smaller than the specific volume of the non-rotator phases which correspond to the liquids and rotator phases. It would appear that the globular molecules in question can turn more easily than one might expect from the free volume $(v - b)$ at their disposal.

Another chemical family with rotator phases which is based on camphor, has also been investigated in detail[7]. D-Camphor has the chemical formula

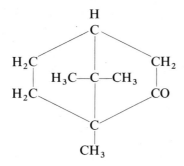

and the molecule would give a sphere by rotation about an axis. This compound is as stable as a plastic crystal over a wide temperature range. At 80°C it has a single relaxation time of $\tau = 11.0 \times 10^{-12}$ sec while at $-20°C$ it has a Cole–Cole distribution with $\alpha = 0.93$.

The activation energy of the relaxation time increases with decreasing temperature, while the dipole moment, derived from the Onsager equation (8.21) decreases with decreasing temperature. Clemett and Davies[7] consider that these two features are linked together, insofar as at lower temperatures the electrostatic interaction of dipoles locks some of them in anti-parallel positions, while the formation of larger aggregates increases the activation energy. They attempt a quantitative interpretation of this mechanism as an order disorder phenomenon (see Chapters 13 and 16), defining a degree of order x according to the Bragg–Williams approximation. This is derived from the measured dipole moment as the fraction of inactive dipoles (the moment of the C=O group being known). The degree of order x is then assumed to enter into the relaxation time according to

$$\tau = A\, e^{\frac{xa}{kT}} \qquad (12.2)$$

where A and a are constant. This treatment is interesting, and gives qualitatively the right answer. Quantitatively it is not very successful, but the method may be capable of refinement.

Other globular molecules as well as certain disc-shaped molecules also show plastic phases. The evidence on these compounds up to 1956 has been summarized by Dryden and Meakins[8].

Long-chain crystals

A second group of dielectrics with plastic solid phases is based on the long-chain paraffins. This group is of great technological interest since several important industrial plastics consist of paraffins or paraffin derivatives with extremely long chains. However, for this group molecular mobility is a more complex phenomenon than for globular compounds.

A normal paraffin has the general formula $CH_3(CH_2)_{n-2}CH_3$ where the number n may vary from zero to several thousand. Paraffins with $n < 6$ or so do not qualify as long-chain paraffins, while paraffins with $n > 100$ or so cannot be produced as pure chemicals. However, a large number of normal paraffins and simple paraffin derivatives with $6 > n < 40$ have been produced chemically pure. Several of these have plastic forms and the study of molecular mobility in crystalline long-chain paraffins is greatly helped by the fact that properties change systematically with chain lengths[9].

A long-chain paraffin in the absence of thermal vibrations has a flat zig-zag shape. Figure 2 shows the crystal structure of polyethylene[10] in terms of the electron densities in the crystalline regions. Figure 2(b) shows the cross-section of the molecule along the plane of the symbolic C—C—C zig-zag. Figure 2(a) shows the cross-section of the crystal structure normal to the chain axes, the central carbon atom being projected on to the plane of the four outside ones. It may be seen that cross-sections form a distorted hexagon.

Figure 2 shows two salient points: firstly, the cross-section of the molecule is approximately cylindrical, and secondly, there are large regions of low electron density between neighbouring molecules. The latter point is understandable when one considers that the electron clouds of paraffins are similar to those of noble gases. Müller[11] has shown that the energy of bonding of paraffins can be theoretically calculated on the basis of dispersion forces, i.e. of transient dipole moments induced by the molecules in their neighbours. These forces are not directional, or only slightly so, and the bonding puts only slight obstacles into the way of certain movements of chains relative to one another.

With regard to the possible movement of paraffin chains it is worth noting that in the gas or liquid these chains are able to twist by rotations, or more exactly step-wise turns, about the C—C bonds of the zig-zag[9]. This is shown by the dielectric behaviour of two diketones[12], with dipole attached in one case at the same side of the chain, in the other at opposite sides. If these molecules remained rigid in the liquid state the dielectric constant of the former ketone would be much larger. However, the dielectric constants are practically equal, indicating extensive twisting over a length of only four or five links of the chain.

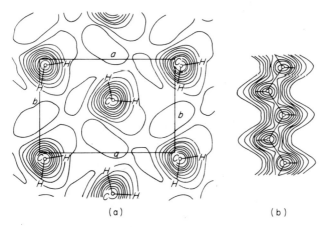

FIG. 2. Electron density maps for polyethylene (from Bunn[10]). (a) In plane perpendicular to c axis at level of carbon atoms; (b) Section through a molecule.

In the solid state, some normal long-chain paraffins occur in phases which are analogous to the rotator phases of the substituted methanes, insofar as they have low heats of melting, etc. These phases are hexagonal, as corresponds to the closest packing of cylindrical rods. It is somewhat confusing that the "plastic" properties of long-chain paraffins in the dielectric context refer only rarely to these phases, and usually to other, more ordered and less symmetrical ones. This is so presumably because the hexagonal phases have only a short temperature range, just below the melting point.

The most usual structure in paraffins is orthorhombic, similar to that shown in Fig. 2, and this is stable over wide temperature ranges. Other forms, still less symmetrical, also occur[9]. Dielectric data on molecular mobility in paraffins were assembled by dissolving dipolar substituted paraffins in normal paraffins or in paraffin wax, the latter being a mixture of normal paraffins of different values of n. The dipolar molecules were often ketones of general formula $CH_3(CH_2)_p CO(CH_2)_q CH_3$, where p and q were usually not very different. Ethers and esters containing the C—O—C group in the chain were also used. Dilute solutions of these molecules give a dielectric constant indicative of free movement of the dipoles about the long chain axes.

Figure 3 shows dielectric results obtained for dilute solutions of long-chain ketones in pure long-chain paraffins of $n = 26$ and $n = 28$ respectively, and in paraffin wax. The relaxation times depend in a regular way on the chain

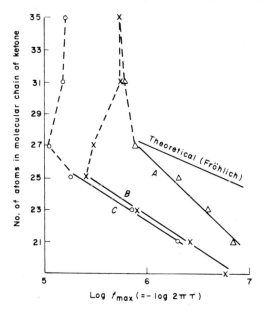

FIG. 3. Variation of relaxation time at 20°C as a function of molecular chain length of long-chain ketones in 0·1 M solution in hydrocarbons. Solvent: Curve A—paraffin wax; Curve B—n-hexacosane ($C_{26}H_{54}$); Curve C—n-octacosane ($C_{28}H_{58}$) (from Meakins[13]).

length as long as the solute chain is shorter than the solvent chains. With solute molecules which are longer than solvent ones this regularity is destroyed, but the dipoles remain mobile.

The evidence on dipolar molecules dissolved in paraffins is discussed in detail by Meakins[13]. It fits the explanation given by Fröhlich[14], which is illustrated in Fig. 4. A dipole (shown as arrow) may reverse its direction if the molecule carrying it performs a "sliding turn". If the chain moved as a rigid

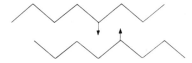

FIG. 4. The "sliding turn" which results in a shift of the molecule and turn of the dipole by 180° (from Fröhlich[14]).

body, this movement would have an activation energy which would increase linearly with the chain length of the dipolar molecule. However, if the chain twists it can wriggle from one position to the other with a smaller energy. For very long chains the energy of turning will then reach a limiting value, since the distant parts of the chain no longer have any effect. Fröhlich gives a formula based on this idea, namely

$$\log 2\pi\tau = A + \frac{B}{T}\tanh\frac{n}{C} \qquad (12.3)$$

where A, B and C are constants and B may be estimated on the basis of the energy needed to twist the chain. This formula gives good agreement with early results; its limitations are discussed by Meakins[13].

The explanation by way of the "sliding turn" is confirmed by the fact that a 5% solution of $C_{11}H_{23}COC_{11}H_{23}$ in $C_{23}H_{48}$ gives no dielectric loss[13] indicating that the ketone chains cannot slide since they are of the same length as the solvent chains. This argument is not quite conclusive, since dipolar mobility may occur in ketone–paraffin mixtures at high temperatures and freeze in at low ones, due to a structural change[9], but it is very strong evidence in favour of the "sliding turn".

The investigations described above refer to molecular mobility in normal long-chain paraffins, revealed by means of dipolar solutes. Pure compounds also show dielectric absorption in β phases. Meakins and co-workers investigated esters of the form $CH_3(CH_2)_pOCO(CH_2)_qCH_3$ where q was zero or small. At temperatures near room temperature these compounds show[15a] an absorption region around 10^7 c/s and another one around 10^{10} c/s, i.e. in the region characteristic for liquids. The latter may be due to movements of the dipole which are nearly independent of the chain as a whole.

Long-chain esters also have "expanded" phases analogous to rotator phases in globular compounds. In these the main absorption is in the 10^{10} c/s region but there are also absorption peaks at very low frequencies. One of the compounds in question forms several α phases, the one nearest to the melting point being hexagonal. Figure 5, according to Dryden[15], shows the complicated behaviour of this compound as a function of temperature. The activation energy in the hexagonal phase is 9 kcal/mole, for the high frequency peak; this phase also has another peak around 10^3 c/s, not shown in the figure. Processes with long relaxation times are found also in expanded phases of acetates and ethers; they have large activation energies which increase linearly with chains length[13]. Meakins[13] gives a table of the activation energies and pre-exponential factors for a wide range of compounds in different structural forms.

The available evidence on expanded or α phases refers to long chain compounds with complicating features. In particular, a polar group near a chain end causes interaction between neighbouring layers of chains[9]. The ester group C—O—CO—C is not a simple dipole, and the C—O—C linkage alters the paraffin backbone. The molecular mobility in a hexagonal expanded phase of a normal long-chain paraffin might be obtained by dissolving a ketone of equal chain length in it, and such a system might perhaps show less complicated relaxation behaviour.

Incidentally, hexagonal expanded phases may be prepared as a mixture of several ketones of equal chain length, but different position of the ketogroup

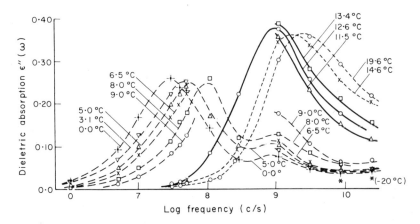

FIG. 5. The dielectric absorption $\varepsilon''(\omega)$ for the α_1, α_2 and α_3 phases of butyl stearate (from Dryden[15]). α_1 ···; α_2 – –; α_3 ——.

"equivalent" ketones[16]. A pure solid ketone shows no dipolar mobility because of the powerful electrostatic interaction of the C=O dipoles, but in a mixture as described the entropy of mixing stabilizes a structure where dipoles are dispersed at different levels[16]. The properties of mixtures of equivalent ketones may be varied in smaller steps than can be done in a study of related compounds, since the number of ketones mixed as well as p and q in $CH_3(CH_2)_p CO(CH_2)_q CH_3$ can be varied independently, while the hexagonal paraffin structure is retained. This might offer good facilities for correlating electrostatic interaction and relaxation times as suggested for the camphors[7], particularly[16a] since structural changes occur with convenient speeds.

Occlusion compounds

In a plastic solid each molecule is more or less free to reorient within a space circumscribed by its nearest neighbours. However, it would be unreasonable to say that the molecule is free to move within a cage, since the neighbours of the molecule are just as free to move as the molecule in question. In occlusion of clathrate compounds individual molecules are contained in holes within a fairly rigid structure, and this facilitates the theoretical analysis of their movements.

Globular molecules like CH_4 have been occluded in β quinol, which contains holes of diameter about 8 Å able to hold one occluded molecule. Thermodynamic examination[17] shows that for methane all rotational and

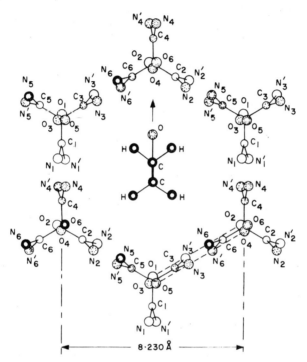

FIG. 6. A tunnel in the lattice of urea (NH_2CONH_2) with an occluded long-chain ketone molecule. The molecules of urea are bonded to each other by hydrogen bonds, indicated by dotted lines in the bottom right-hand corner of the "cage", formed by urea (from Lauritzen[19]).

vibrational degrees of freedom are active above about 100°K. The dielectric relaxation of some polar compounds occluded in quinol was found[13, 18] to give perfect Debye curves; at room temperature the maxima were mostly at microwave frequencies. The energy of activation was 2·3 kcal/mole for occluded methanol. For the methyl cyanide complex the activation energy is much higher, 18 kcal/mole, i.e. the molecule is held more tightly. It was found that the cyanide molecule distorts the cage[18]. The activation energy of the methanol complex is higher than that of the solid methyl halides in Table I, which is interesting if the discrepancy is not due to hydrogen bonding between methanol and the quinol cage.

Long-chain paraffins and their derivatives also form occlusion compounds, since they can be accommodated in straight tunnels within the lattice of urea. Figure 6 shows a cross-section of a tunnel containing a ketone molecule. The edge of the hexagon is not much larger than the corresponding spacing in a hexagonal paraffin lattice.

The dielectric relaxation of several ketones and other substituted paraffins occluded in urea has been investigated[20]. The evidence at room temperature indicates extensive twisting of the chains, in that the relaxation time is nearly

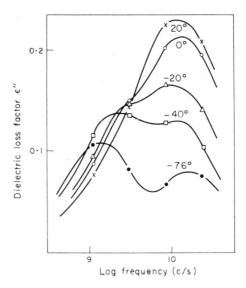

FIG. 7. The dielectric absorption of the ketone 16-hentatriacontanone ($C_{15}H_{31}COC_{15}H_{31}$) occluded in urea, as a function of frequency and temperature (from Meakins[20]).

independent of chain length, and a doubly substituted derivative[12], namely 1:10 dibromedecane, gives a value of ε_s corresponding to independent movement of the two dipoles. The distribution of relaxation times is very temperature dependent, as shown[20] in Fig. 7.

The behaviour of the occlusion compounds illustrated in Figs 6 and 7 may be explained theoretically by considering the force field within which the dipole moves. The plane projection shown in Fig. 6 appears to allow six positions of equal potential energy for the C=O dipole. However, consideration[19] of the structure of urea in three dimensions indicates that there should be six potential minima which differ from each other.

This consideration[19] assumes that the ketone molecule rotates about its axis. Lauritzen calculates the relaxation of the model by a method developed by Hoffmann[21,22]. He defines occupation numbers $N_1, N_2, \ldots, N_6$ for the six sites, and six different equations

$$\frac{dN_1}{dt} = 2k_1N_1 + k_6(N_2 + N_6)$$

$$\vdots \qquad (12.4)$$

$$\frac{dN_6}{dt} = -(k_2 + k_6)N_6 + k_1N_1 + k_5N_5$$

containing seven constant coefficients k_1 to k_7.

For a model with six different potential wells in a ring the calculation leads to five time constants. The model illustrated gives only four time constants because well 2 is equal to well 6 and well 3 is equal to well 5.

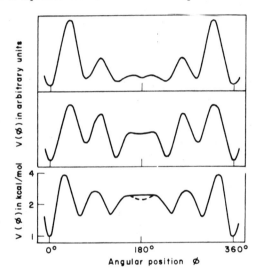

FIG. 8. The potential energy as a function of angular position for 16-hentatriacontanone in urea. The upper two curves are calculated (using 6–12 potential and volume overlap, respectively) and the lowest one is derived from Meakins[20] dielectric data (from Lauritzen[19]).

The depth of the wells and hence the constants k_1, k_2, etc. can be calculated on the basis of the forces between the occluded molecule and the molecules of the cage. Lauritzen finds the potential energy as a function of angle theoretically and compares it with the potential energy function which explains the dielectric data. Figure 8 shows the comparison with theoretical data referring to two methods of calculation. The agreement is not perfect and this is not surprising when one bears in mind that Lauritzen assumes the chains to rotate about their axes, while experiment shows that extensive twisting occurs. However, the example given is one of the rare cases where a distribution of relaxation times could be derived with fair precision on the basis of intermolecular forces.

B. IONIC CRYSTALS

An ideal ionic crystal may be exemplified by a perfect crystal of NaCl, where each Na^+ ion is surrounded by six Cl^- ions and *vice versa*. Such a crystal can be polarized only by way of a perturbation of its thermal vibrations. In practice, all crystals contain dislocations and besides, the applications of metallic electrodes to the surface of even an initially perfect ionic crystal causes complications which will be discussed in Chapter 14. The

Na^+	Cl^-	Na^+	Cl^-	Na^+	Cl^-
Cl^-	Ca^{++}	Cl^-	Na^+	Cl^-	Na^+
Na^+	Cl^-		Cl^-	Na^+	Cl^-
Cl^-	Na^+	Cl^-	Na^+	Cl^-	Na^+

FIG. 9. A 001 face of an NaCl crystal, where one sodium ion is replaced by calcium.

present section is concerned with imperfections within the bulk of ionic crystals, i.e. it is not concerned with electrode effects.

The most straightforward way of introducing[23] a polarizable flaw into an ionic crystal, say NaCl, is to replace an ion by another ionic species of different valency, say Na^+ by Ca^{++}. This can be done without distorting the lattice appreciably, as long as the ionic radii are sufficiently similar. The introduction of Ca^{++} is possible in the bulk of the crystal only if neutrality is maintained, i.e. if one Ca^{++} ion replaces two Na^+ ions. Thus the simplest imperfection in question consists of a Ca^{++} ion and a vacant site. Since the two elements of the pair in question are oppositely charged, they attract each other, and there is thus a high probability that they will remain in close proximity, as shown in Fig. 9. The pair shown constitutes a mobile dipole, since a vacant site may be filled by any adjacent positive ion.

However, the mechanism of dielectric relaxation which follows from the presence of ion-vacancy pairs is complicated by the presence of electrical conductivity. Firstly, there is always a finite probability that ion and vacancy will become separated, and that either or both may migrate over many atomic distances. This implies finite ionic conductivity. Secondly, there is a possibility that electronic conduction may occur.

Relaxation due to small ionic displacements

Many imperfect ionic crystals show relaxation effects which are only slightly perturbed by direct conductivity, electronic or ionic. For some of

12. DIELECTRIC RELAXATION IN CRYSTALLINE SOLIDS

these dielectric relaxation can be studied with accuracy and interpreted in terms of structure.

A great deal of information is available on alkali halides with divalent cations as impurities[13,24,25]. For these systems τ tends to be rather long at room temperature (it may be as long as 1 sec), but at high temperatures $\varepsilon''(\omega)$ may be followed conveniently and indicates relaxation times which are single within the limits of the experimental error, set largely by incidence of ionic conductivity. The relaxation times obey[13,25] equation (4.47) where A is $2\text{--}4 \times 10^{-14}$ sec while W is of the order 12 kcal/mole, values which are plausible for a fairly fast process of diffusion in a solid.

The relaxation of a cation-vacancy dipole has been treated theoretically and experimentally by Wachtman[26], both with regard to dielectric and mechanical response. The treatment refers to ThO_2, containing small amounts of CaO. The ThO_2 lattice is cubic, each anion having eight oxygen ions as neighbours. The introduction of CaO has the effect of replacing a Th^{++++} ion by a Ca^{++} ion, while one oxygen is missing and the resulting vacancy is tightly bound to the Ca^{++} ion by Coulomb forces.

A model with eight alternative dipole directions might be treated by Hoffmann's method (see equation 12.4). However, the eight positions in the present case are arranged in three dimensions, while the six positions entering into equation (12.4) are in a plane, and the modes of relaxation for the three-dimensional cases are more complicated. Wachtman uses group theory to classify eight relaxation modes. The mathematical formalism involved has been recently reviewed[27].

The most interesting result of Wachtman's treatment is that mechanical and electrical relaxation do not refer to the same relaxation modes. An ultrasonic vibration affects the configuration of the ion-vacancy dipole by way of a shear stress, while an alternating electrical field affects tensile relaxation modes which are unaffected in the mechanical case.

The difference between the response to mechanical and electrical perturbation has the consequence that the mechanical relaxation time should be half the electrical one, neglecting electrostatic interaction. This signifies that the shear modes of the spontaneous fluctuations of the polarization relax twice as fast as the tensile ones, and that ultrasonic and electrical measurements determine respectively the relaxation times of these two types of fluctuations. The experimental data confirm the theoretical ratio of the two relaxation times, but Wachtman suggests further experiments on the internal friction of single crystals which would show whether the mechanical modes are indeed unaffected by tensile stress. The conductivity of ThO_2 is very small and electronic rather than ionic.

Ionic movements in crystals of α quartz (SiO_2) and their relationship to dielectric relaxation have been studied very thoroughly. Stevels and Volger[28], in one of a long series of papers on related subjects, report on dielectric relaxation at low temperature in α quartz which contains aluminium. The Al^{+++} ion occupies an Si^{++++} site, and charge neutrality is preserved by incorporated alkali ions in interstitial positions.

Figure 10 shows that an alkali ion has more than one position of minimum potential energy near the aluminium ion whose charge deficiency it compensates. The situation can be symbolized by a potential well model as shown,

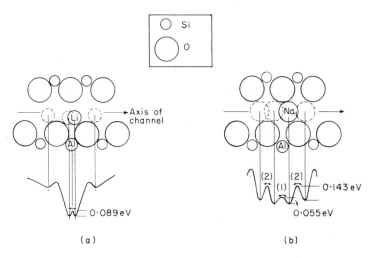

FIG. 10. The positions for alkali ions in the channels of α quartz, and corresponding schematic curves for the potential energy as a function of distance along the channel (from Stevels and Volger[28]).

distance being counted along a channel present in the perfect crystal. Dielectric relaxation at low temperatures corresponds to the fluctuation of the position of the alkali ion between the pairs of sites of lowest potential energy. The activation energy of this process is small, as indicated in the figure; the pre-exponential factor A in equation (4.47) is normal, 2×10^{-13} for Na as well as Li.

Thermally activated charge carriers and their role in conduction and relaxation

The argument so far refers to ions whose movements are restricted to displacements over atomic distances. These small movements are thermally activated in the sense that equation (4.47) or (4.48) applies. This means that the movements of the ion can be represented by a potential well model where the ion jumps from site to neighbouring site. For jumps between neighbouring sites the bistable model applies. According to Chapter 4, Section D the relaxation time of the bistable model is $\tau = 1/2w$, and if equation (4.47) applies:

$$w = \frac{1}{A} e^{-\frac{W}{kT}}. \tag{12.5}$$

If many accessible sites are available the bistable model can easily be adapted to a model of ionic conductivity. We shall consider a large number N of potential wells in a simple cubic array. The distance between neighbouring wells is a and the potential hill separating neighbouring wells has a height W. A number $n \ll N$ of particles jump from well to well in a random fashion, defined by equation (12.5). A similar model for ice is discussed in Chapter 15, Section B.

In thermodynamic equilibrium the concentration $c = n/N$ of occupied wells will be constant for this model, independent of spatial coordinates. However, if an external agency causes a departure from equilibrium so that at a given time the concentration depends on x, then the concentration differences will subsequently equalize according to Fick's equation:

$$\frac{\partial c}{\partial t} = D \frac{\partial^2 c}{\partial x^2} \tag{12.6}$$

where D is the coefficient of diffusion, which is constant for the random jumping assumed. It can be simply shown (see, for instance, Mott and Gurney[29] section IIs; Dekker[30] sections 3.5 and 7.5) that for the model assumed here

$$D = \frac{1}{6} \cdot \frac{n}{N} a^2 w. \tag{12.7}$$

(The factor $\frac{1}{6}$ corresponds to a simple cubic lattice. A factor $\frac{1}{3}$ would apply to the self-diffusion of Na in NaCl.)

Equation (12.7) applies to a model where the influence of electrical charge is neglected. However, this influence is important. Figure 10 shows that Na^+ or Li^+ in α quartz is bound to sites near Al^{+++} by Coulomb forces which contribute to the depth of the deep wells. A simple model such as that which gives equation (12.7) would here signify that alkali ions can jump from oxygen to oxygen undisturbed by the Coulomb forces which cause deep wells. If aluminium ions are far apart, we can envisage diffusion along long stretches of the channel provided that alkali ions can escape from the deep wells. Their escape is governed by a dissociation equilibrium which may most simply be written

$$Na_2O \rightleftharpoons 2\,Na^+ + O^{--}. \tag{12.8}$$

In this case the coefficient of diffusion of Na$^+$ ions would depend not only on the number of successful jumps w of escaped ions but also on the number of free ions. The latter would be governed by the dissociation equilibrium.

The diffusion of ions in real crystals is complicated not only by Coulomb forces and the circumstance that usually more than one charge carrier is mobile, but also by textural features which facilitate some types of movement. Ionic conductivity has been reviewed by Lidiard[31], who also discusses expressions for the coefficient of diffusion in some detail.

While observed coefficients of diffusion cannot generally be calculated from available structural data, a general relationship between the coefficient of diffusion D and the ionic mobility μ hold over a wide range of conditions[31]. This is the Einstein or Einstein–Nernst relation which may be derived as follows[29,31]. We assume $n(x)$ ions of charge e/unit volume, in a field which varies only in the x direction, so that

$$E = -\frac{\partial V}{\partial x}. \tag{12.9}$$

According to Boltzmann statistics we have in equilibrium

$$n(x) = \text{const.} \ e^{-\frac{eV}{kT}}. \tag{12.10}$$

In equilibrium no current may flow. This means that as many ions as flow in the field direction as a consequence of the applied field must flow in the opposite direction as a consequence of diffusion. Diffusion tends to equalize the concentration differences caused by the transport of matter under the influence of the electric field. Zero current signifies

$$\mu n e E - e D \frac{\partial n}{\partial x} = 0 \tag{12.11}$$

where μ is the mobility. Equation (12.11) is satisfied if

$$n(x) = \text{const.} \ e^{-\frac{\mu V}{D}}. \tag{12.12}$$

Comparison of equations (12.10) and (12.12) shows that

$$\frac{\mu}{D} = \frac{e}{kT} \tag{12.13a}$$

which is the Einstein relation. The charge in equation (12.13) is in e.s.u. It is often convenient to use practical units, when

$$\frac{\mu}{D} = \frac{96{,}500}{RT} \tag{12.13b}$$

12. DIELECTRIC RELAXATION IN CRYSTALLINE SOLIDS

R being the gas constant and $eN = 96{,}500$ the charge in Coulombs associated with Avogadro's number of electrons. The volume conductivity is given by

$$\gamma = ne\mu \qquad (12.14)$$

which defines the "drift" mobility μ.

Equations (12.5), (12.7) and (12.13) show the significance of thermal activation for relaxation and conduction. The term w in equation (12.5) defines the relaxation time of random hopping between two wells, and it enters as a factor into D and μ. The frequency of jumps w is generally not the only temperature dependent term in μ, but it is exponential and μ increases with temperature as long as $W \gg kT$. The relaxation time of the bistable model $\tau = 1/2w$ is not the only relaxation time which may apply to a system of many potential wells. Relaxation times for a model consisting of a linear chain of potential wells, of limited lengths, are defined in Chapter 15, Section A. A dissociation equilibrium as in (12.8) has a relaxation time which may be relevant for dielectric relaxation as discussed in Chapters 11, 14 and 15.

"Hopping" Electrons

Electronic conduction differs in general from ionic conduction, insofar as electrons may be treated in terms of the band model. In the band model an electron is characterized by an energy level and other parameters which are attributes of the solid as a whole, not of a distinguishable, individual particle. Conduction electrons in the sense of the band model have mobilities which are normally large compared with $\mu = 1$ cm^2 sec^{-1}, and decrease with rising temperature. The drift mobility defined by equation (12.14) is not the only kind of mobility, and the theory of semi-conductors is a large subject. The decrease of the mobility with rising temperature is explained by interactions between electrons and lattice vibrations[32].

Charge transport by electrons which behave according to the band model may lead to dielectric relaxation in inhomogeneous dielectrics, as discussed in Chapter 14, Section A. This is trivial. However, it has recently been suggested that electrons sometimes cause dielectric relaxation by moving like thermally activated charged particles, i.e. by "hopping".

It is possible to envisage a more or less gradual change from quantum mechanical to classical behaviour in the following sense. Electrons of very low mobility have according to the band model high effective mass, due to the strong interaction between electrons and thermal vibrations. Fröhlich and Sewell[33] point out that mobilities $\mu < 0{\cdot}1$ cm^2 sec^{-1} (if they are genuine lattice mobilities) contradict basic conditions for the validity of the band model.

Insofar as the band model breaks down hopping seems reasonable, but the quantitative formulation of such behaviour raises difficult problems[34]. Fröhlich points out that hopping electrons would give rise to relaxation effects[34].

Hopping electrons have been investigated experimentally mainly from the viewpoint of conduction. Mobilities which increase with temperature as in thermally activated processes were found by Morin[35] in a classical paper which refers to NiO doped with Li^+ or Cr^{+++}. The number of carriers was derived in this case by an ingenious combination of physical and chemical methods.

Mechanical relaxation effects in Li doped NiO were found[36] by van Houten and interpreted in terms of the hopping model. Ultrasonic data are stronger evidence than dielectric data in this context, since they are less subject to disturbing effects (see Chapter 14, Sections A and C). Nevertheless, the evidence for hopping conduction and even more for relaxation is still very controversial. It refers mainly to transition metal oxides which are often not stoichiometric and whose chemical physics is extremely complicated[37]. The uncertainties about hopping conduction are brought out in the informal report of a recent conference[38].

While the above discussion refers mainly to fairly conductive crystalline materials, relaxation effects have been attributed also to electrons in crystals which are good dielectrics. Volger[25] reports relaxation in α quartz at low temperatures. Irradiation with electrons or X-rays produces relaxation peaks which are attributed to electrons. In one case the activation energy is 6.6×10^{-3} eV, with A in equation (4.47) equal 6×10^{-7}, an unusually high value.

REFERENCES

1. J. Timmermans (1961). *Physics Chem. Solids*, **18**, 1.
2. W. J. Dunning (1961). *Physics Chem. Solids*, **18**, 21.
3. W. C. Price and G. R. Wilkinson (1961). *Physics Chem. Solids*, **18**, 74.
4. C. P. Smyth (1961). *Physics Chem. Solids*, **18**, 40.
5. C. Clemett and M. Davies (1962). *Trans. Faraday Soc.*, **58**, 1705.
6. S. Glasstone, K. J. Laidler and H. Eyring (1961). "Theory of Rate Processes", p. 487. McGraw Hill, New York.
7. C. Clemett and M. Davies (1962). *Trans. Faraday Soc.*, **58**, 1718.
8. J. S. Dryden and R. J. Meakins (1957). *Rev. pure appl. Chem.*, 7, 15.
9. V. Daniel (1953). *Adv. Phys.*, **2**, 450.
10. C. W. Bunn (1939). *Trans. Faraday Soc.*, **35**, 482.
11. A. Müller (1936). *Proc. R. Soc.*, A, **154**, 624.
12. A. Müller (1940). *Proc. R. Soc.*, A, **174**, 137.
13. R. J. Meakins (1961). *Progr. Dielect.*, 3, 151.
14. H. Fröhlich (1958). "Theory of Dielectrics." Oxford University Press, London.
15. J. S. Dryden (1957). *J. Chem. Phys.*, **26**, 604.

15a. J. S. Cook and R. J. Meakins (1959). *J. Chem. Phys.*, **30**, 787.
16. V. Daniel (1950). *Phil. Mag.*, **41**, 631.
16a. V. Daniel and C. Turner (1953). *Phil. Mag.*, **44**, 1371.
17. L. A. K. Staveley (1961). *J. Phys. Chem. Solids*, **18**, 46.
18. J. S. Dryden (1953). *Trans. Faraday Soc.*, **49**, 1333.
19. J. I. Lauritzen (1958). *J. Chem. Phys.*, **28**, 118.
20. R. J. Meakins (1955). *Trans. Faraday Soc.*, **51**, 953.
21. J. D. Hoffmann and H. G. Pfeiffer (1954). *J. Chem. Phys.*, **22**, 132. Also (1954). *J. Chem. Phys.*, **23**, 1331.
22. J. D. Hoffmann and B. M. Axilrod (1955). *J. Res. natn. Bur. Stand.*, **54**, 357.
23. R. G. Beckenridge (1948). *J. Chem. Phys.*, **16**, 959. Also (1950). *J. Chem. Phys.*, **18**, 913.
24. J. S. Dryden and R. J. Meakins (1957). *Dis. Faraday Soc.*, **23**, 39.
25. J. Volger (1960). "Progress in Semiconductors", Vol. 4, p. 207. Heywood, New York.
26. J. B. Wachtman (1963). *Phys. Rev.*, **131**, 517.
27. A. S. Nowick and W. R. Heller (1965). *Adv. Phys.*, **14**, 101.
28. J. M. Stevels and J. Volger (1962). *Philips Res. Rep.*, **17**, 283.
29. N. F. Mott and R. W. Gurney (1940). "Electronic Processes in Ionic Crystals." Oxford University Press, London.
30. A. J. Dekker (1958). "Solid State Physics." Macmillan, London.
31. A. B. Lidiard (1957). "Encyclopedia of Physics." (ed. S. Flugge), Vol. 20, p. 246. Springer-Verlag, Berlin.
32. J. M. Ziman (1960). "Electrons and Phonons." Oxford University Press, London.
33. H. Fröhlich and G. L. Sewell (1959). *Proc. Phys. Soc.*, **74**, 643.
34. H. Fröhlich (1963). E.R.A. Report No. 5003.
35. F. J. Morin (1958). *Bell Syst. tech. J.*, **37**, 1047.
36. S. van Houten (1962). *J. Physics Chem. Solids*, **23**, 1045.
37. F. A. Kröger (1964). "The Chemistry of Imperfect Crystals." North Holland Publications, Amsterdam.
38. E. R. Schatz (1964). "Transition Metal Compounds: Transport and Magnetic Properties." Gordon and Breach, New York and London.

CHAPTER 13

Insulating Materials With Amorphous and Glassy Constituents

Practical solid insulation consists largely of plastics, that is organic polymers and their mixtures or compositions. Also, inorganic insulating materials such as silicate glasses and ceramics are practically useful. The vast majority of the materials (with the exception of a few truly crystalline materials such as mica) have in common that they have amorphous constituents. Besides, many practical materials are used in a state which does not correspond to thermodynamic equilibrium, i.e. they are more or less glassy†.

A. ORGANIC POLYMERS

Organic polymers consist generally of chains whose links are characteristic groups, such as CH_2. These chains may either be separate macromolecules, or they may be more or less cross linked by chemical bonds and thus form a three-dimensional network. A general formula, which fits a large number of polymer chains, may be written in the form

$$\begin{array}{cccc} R_1 & R_4 & R_1 & R_4 \\ | & | & | & | \\ -C\!-\!\!&\!\!C\!-\!\!&\!\!C\!-\!\!&\!\!C- \\ | & | & | & | \\ R_2 & R_3 & R_2 & R_3 \end{array}$$
(I)

where R_1 to R_4 are molecular groups. The unit sketched is repeated in a chain some 10–1000 times, and a practical polymer is a mixture of chains of different length.

Table I shows dielectric and other data[1,2] for several polymers which can be represented by formula I, as well as for polychlorisoprene of formula

$$\begin{array}{cccc} H & H & Cl & H \\ | & | & | & | \\ -C\!-\!\!&\!\!C\!=\!\!&\!\!C\!-\!\!&\!\!C- \\ | & & & | \\ H & & & H \end{array}$$
(II)

for the unit of the chain and cellulose of formula

† The present review of this group of materials is relatively short and is intended as an introduction to the basic ideas in this field. The subject is more fully covered in "Anelastic and Dielectric Effects in Polymeric Solids" by N. G. McCrum, B. E. Read and G. Williams, John Wiley, 1967.

TABLE I

Material	R_1	R_2	R_3	R_4	ε		$\tan\delta \times 10^4$		Maximum working temperature °C	Type	Remarks
					10^2 c/s	10^8 c/s	10^2 c/s	10^8 c/s			
Polyethylene	H	H	H	H	2·3	2·3	0·5–1	1–2	70–110	A	Tough, can be drawn
Polytetra-fluoroethylene	F	F	F	F	2·2	2·2	2	5	~250	A	Hard, rigid
Polyisobutylene	H	H	CH_3	CH_3	2·2	2·2	2	5		A	Rubbery, soft
Polystyrene	H	H	H	—C$_6$H$_{11}$	2·6	2·6	2	5	70	A	Tough, strong
Polyacrylonitrile	H	H	H	CN	2·8	2·6	70	70	100	B	Forms fibres
Polyvinylchloride	H	H	H	Cl	3·2	2·8	200	100	80	B	Tough
Polychlortri-fluoroethylene	F	F	F	Cl	2·7	2·3	200	50	200	B	Hard, rigid
Polymethyl-metacrylate	H	H	CH_3	—COCH$_3$ (=O)	3·4	2·6	600	60	80	B	Rigid, fairly tough
Polychlorisoprene	see formula II				7	3	10,000	1000	100	C	Rubbery, tough
Cellulose	see formula III				7·5	5·5	100	700	90	D	Main constituent of paper

From Garton[5].

13. INSULATING MATERIALS WITH AMORPHOUS AND GLASSY CONSTITUENTS

$$\begin{array}{c}
H_2C-O-H \\
| \\
C-O \\
H \\
-CH \qquad\qquad HC-O- \\
H\ H \\
C-C \\
|\ | \\
O\ O \\
|\ | \\
H\ H
\end{array}$$

(III)

for a half unit of a chain.

The electrical data refer to typical pure and dry materials which are commercially available, and to room temperature. Two values are given for dielectric constant and loss, one for 100–1000 c/s, the other for the region around 10^8 c/s. The low frequency values include in general some contribution from a d.c. conductivity, which may be very sensitive to moisture and impurities. In consequence, the figures quoted should not be taken as standard values. However, the data may serve as a guide for classifying the materials into four groups. The table also gives an approximate limit for the temperature up to which the materials may be used in practice.

The table represents only a minute fraction of the vast experimental material available. Curtis[1] gives a systematic review of the dielectric properties of organic polymers. Dielectric[2] and mechanical[3] data are reviewed within two symposia edited by von Hippel, which also contain useful tables, while further data appear in books on electrical insulation[4] and in sources enumerated in Appendix VI.

The division into types which the table illustrates[5] may be justified as follows. Type A are polymers without permanent dipoles, with low dielectric loss which varies only little over a wide frequency range. Type B contain permanent dipoles, but the chains are at room temperature and mostly immobile over periods of the order of a second. Nevertheless, these polymers contain mobile dipoles, as indicated by the loss angle. Type C polymers are rubbery at room temperature, extensively disordered, and the chains and most dipoles are more or less mobile. If polyisobutylene contained dipoles it would belong to Type C rather than Type A. Cellulose might be in Type B, but has been classified as a separate type because of its extensive hydrogen bonding which causes complications, and this will be discussed in Chapter 15.

Even the short selection of data quoted shows that mechanical properties may differ greatly as between polymers whose dielectric properties are

Non-polar polymers and the problem of $\frac{1}{f}$ noise

Table I shows that the dielectric loss is lowest for non-polar polymers. There are many practical applications where low loss is vital. The stringent requirements of transatlantic submarine telephone cables have led to intensive investigations of polyethylene with minimum loss. This work is described in a recent authoritative survey[9].

Reddish[9] and co-workers describe very accurate measurements on low loss polythenes of different purity and density over a wide range of conditions. For one particular material tan δ varies between about 2×10^{-5} and 1×10^{-4} over the temperature range $-120°C < T < 60°C$ and frequency range $10^2 < \omega < 10^7$. The contour map showing these data contains peaks which are analysed in terms of loss mechanisms. The γ mechanism is attributed to some movement which requires at least four CH_2 groups in line, that is to a movement of the polymethylene chains. The loss due to this process increases with frequency. The authors conclude that attempts to reduce the loss of polyethylene further are governed by the law of diminishing returns.

The results for polythene suggest that there is a minimum loss that must be present even in the purest non-polar material. This may perhaps be understood, at least to some extent, in terms of second-order dipole moments[10]. The interaction of two thermal vibrations may produce a distortion of even a normally symmetrical electron cloud. This distortion signifies a dipole whose relaxation time will be related to the sum or difference of the frequencies of the two vibrations. The relaxation of such dipoles can in principle extend down to very low frequencies. However, polyethylene cannot be treated as a strictly non-polar molecule[9].

Even though peaks may be discerned in the dielectric loss of polyethylene $\varepsilon''(\omega)$ is yet remarkably constant over a wide frequency range. As shown in Chapter 7 dielectric loss of this kind corresponds to a distribution function $y(\tau)$ given by equation (7.18).

Substantially constant loss over a wide frequency range is common to many low loss materials. It is a phenomenon which is related to the so-called $1/f$ noise which is observed in many physical situations. This may be seen by calculating the real $\varepsilon'(\omega)$ corresponding to equation (7.18) according to

13. INSULATING MATERIALS WITH AMORPHOUS AND GLASSY CONSTITUENTS

equation (5.13) as

$$\varepsilon'(\omega) = \varepsilon_\infty + \frac{\text{const.}}{\omega} \left| \arctan \right|_{\tau_a}^{\tau_b} \tag{13.1}$$

where τ_a and τ_b are the limits over which $y(\tau)$ is constant. If the frequency range is wide the arc tan term may be put equal $\pi/2$.

Equation (13.1) implies that $\varepsilon'(\omega)$ depends inversely on frequency. Now, according to Chapter 4, $\varepsilon'(\omega)$ characterizes the magnitude of the fluctuation of the polarization in a given frequency range around ω. However, this quantity is related to the noise which may be measured as the mean square value of the fluctuations of the voltage which occur spontaneously between electrodes applied to the dielectric (see equation 4.40). Hence, the distribution function (7.18) characterizes $1/f$ noise.

In a discussion of $1/f$ noise in semi-conductors Sautter[11] states that such noise is found over a frequency range of eight decades. He discusses various models to explain the occurrence of this noise spectrum. A recent model[12], not very successful quantitatively, has the merit that it can be simply explained in physical terms. The unlimited $1/f$ spectrum is taken to correspond to the unlimited delays which occur when a group of carriers in the conduction band has random access to an equal number of recombination centres.

As for dielectrics, Garton[13] draws attention to the fact that $y(\tau)$ is the product of the number of polarizable elements and their polarizability (equation 5.8). He considers a model where the polarizable elements are dipoles consisting of double potential wells where one well is very much deeper than the other, in the absence of a field. The relaxation time of such a dipole can be calculated in terms of the relaxation times τ_1 and τ_2 of both wells, where τ_1 is the relaxation time of a symmetrical dipole characterized by the energy difference between well 1 and the potential hill and τ_2 similarly characterized by the depth of well 2. The relaxation time of the unsymmetrical combination is

$$\tau = \frac{2\tau_1 \tau_2}{\tau_1 + \tau_2} \tag{13.2}$$

while the polarizability of the combination is

$$\mu = \text{const.} \frac{\tau_1 \tau_2}{kT(\tau_1 + \tau_2)^2}. \tag{13.3}$$

This model leads to an exactly frequency-independent $\varepsilon''(\omega)$ over a frequency range which may be chosen as wide as desired. The derivation of this result is based on the assumption that all deep wells have equal time constants τ_2, but that the shallow wells are thermally activated. This implies that the number of shallow wells with activation energies within an internal $W \pm dW$ is

$$dN = e^{-\frac{W}{kT}} \times \frac{dW}{kT} \qquad (13.4)$$

while the relaxation time of a shallow well is assumed to be determined by the same energy W according to equation (4.47). The simultaneous validity of equations (13.4) and (4.47) implies that the shallow wells are not due to a change in the energy levels of the deep wells, but arise independently of them. Garton's interpretation of the two kinds of well is that deep wells next to shallow ones occur fairly frequently in amorphous materials while they are exceptional in crystalline materials, being restricted to disordered regions, such as grain boundaries. This theory has so far not been experimentally verified. Even the theory of $1/f$ noise in semi-conductors is still controversial, in spite of the wider range of applicable experimental techniques.

Frequency independent loss occurs also in mechanical relaxation. Pelzer[14] discusses mechanical models as well as electrical circuits which lead to nearly frequency independent loss.

Relaxation in polymers with polar group: general

Table I illustrates that polar polymers of type B have only moderate loss angles at room temperature and are mechanically rigid while other polymers, denoted by type C, are soft and have high losses. This description refers to a single temperature. When polymers of type B are examined at different temperatures, they behave in a way exemplified by Fig. 1. With rising temperature the polymer softens and peaks of the mechanical and electrical loss appear in a temperature region which might be considered as a transition between solid and liquid. The polymer may in addition show relaxation peaks at lower temperatures. In the case quoted, two small peaks appear in the mechanical loss, but not in the dielectric one. However, dielectric low temperature loss peaks appear in many polymers. In broad terms the low temperature peaks refer to movements of individual molecular groups, while the high temperature peak corresponds to a cooperative disordering of the solid, a kind of gradual melting.

In the last resort, both high and low temperature relaxation are connected with the extent of ordering of a polymer. Molecular groups, attached to a large molecule, may remain mobile within a crystal, as is shown by Fig. 11.2 for the pure compound lupenone. Besides, Chapter 12 shows that globular

13. INSULATING MATERIALS WITH AMORPHOUS AND GLASSY CONSTITUENTS

molecules are capable of large changes of orientation within perfect crystals. However, it is clear that the presence of disorder increases the possibilities of

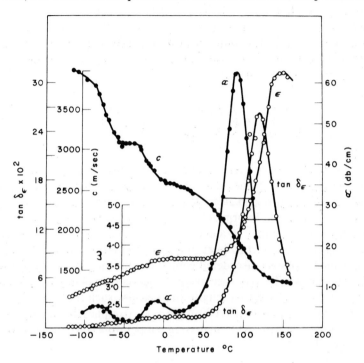

FIG. 1. Comparison between dielectric and mechanical properties of poly(vinyl acetate) at 2 Mc/s. Dielectric constant ε', loss tangent tan δ, and velocity c m sec^{-1} and attenuation α db cm^{-1} of longitudinal ultrasonic waves as functions of temperature. The attenuation α expressed in db cm^{-1} is larger than in nepers cm^{-1} by a factor 8·686 (Thurn and Wolf[15] quoted by Saito et al.[8]).

molecular movements, as may be seen in Chapter 12, Section B for long-chain crystals. Disordered regions are present in even the most orderly polymers and at low temperatures. Polymers of type C are largely disordered at all temperatures.

Order in polymers and its relaxation

The word order, when applied to a polymer, conveys a qualitative meaning but it is not easy to define in terms of one or even of several variables which have an exact meaning. The fundamental approach to the problem of order in polymers is via the partition function and thus the free energy, derived on the basis of all possible molecular configurations and their energies. This approach is covered in recent treatises[6,7,8]. In the present chapter we shall discuss the problem in terms as straightforward as possible, at the risk of some over-simplification.

The simplest, although crude, way of considering order in polymers is to assume that disorder can be defined by a single variable x, which is zero for perfect order and unity for complete disorder, and that the free energy can be written down in terms of this variable. In this case, the equilibrium degree of order follows from the equation

$$\left(\frac{\partial G}{\partial x}\right)_{p,T} = \left(\frac{\partial H(x)}{\partial x}\right)_{p,T} - T\left(\frac{\partial S(x)}{\partial x}\right)_{p,T} = 0 \tag{13.5}$$

where G is the Gibbs free energy and H the enthalpy. Qualitatively, it is clear that H increases with x since the cohesive forces are better satisfied in the crystalline regions. It is also clear that S increases with x and that $\partial S/\partial x$ in a polymer is a large term. In principle, equation (13.5) applies to any solid, but the equilibrium value of x in a crystal is very small because the entropy of disorder plays a small part in determining the free energy. In a polymer there are many more possible disordered configurations than in a solid consisting, say, of hard spheres. This is so because polymers consist of more or less flexible macro-molecules of different length and because such molecules may flex, twist, coil and become entangled. Equation (13.5) implies that the equilibrium degree of order decreases with increasing temperature since the entropy term is multiplied by T.

The above discussion shows that polymers are to some extent disordered in equilibrium. In practice, they are more disordered than corresponds to equilibrium. This is particularly so if a polymer is prepared hot and then chilled to a temperature where structural changes are exceedingly slow.

The process whereby a polymer approaches its equilibrium degree of order is a relaxation process. The relaxation time of this process may be derived using irreversible thermodynamics, if the free energy can be calculated in terms of the degree of order. An analogous treatment was carried out in Chapter 3 for the polarization of the double potential well model, where the order variable ξ was chosen to correspond to a $\frac{1}{2} - x$ in the notation used here, where $x = 0$ corresponds to complete order.

The double potential well is not a very suitable model for the description of the ordering of polymers, but it can serve as an illustration for the way in which ordering may be cooperative. If the dipoles represented by the double well interact with each other electrostatically the free energy of the double well model is given by equation (3.27). The procedures given in Chapter 3 for the derivation of the relaxation time from the free energy (equations 3.31 and 3.33) lead to an expression

13. INSULATING MATERIALS WITH AMORPHOUS AND GLASSY CONSTITUENTS

$$\tau = \tau_0 \frac{T}{T - T_c} \tag{13.6}$$

where τ_0 is the relaxation time in the absence of dipolar interaction and T_c is a constant for a particular configuration of dipoles. The equation signifies that cooperative dipolar interaction lengthens the relaxation time. The polarization of the system is also enhanced according to equation (3.27).

While the above treatment of the double potential well is one of the simplest treatments of the relaxation of a cooperative phenomenon, it is not an exact analogue of the disordering of the polymer. Order in a polymer signifies that macro-molecules are packed so that as few holes as possible are left unfilled and that the maximum possible number of bonds between molecules are made. If order has been achieved, neighbouring molecules hold each other in place, they cooperate to maintain the order. To that extent dipolar interaction and the interaction of molecules are analogous. However, dipolar interaction acts over long distances, while molecules in polymers act only on their nearest neighbours. It is possible to treat order–disorder phenomena based on nearest neighbour interaction theoretically. The subject has been surveyed by Temperley[16] in a general way.

Order and the processes of ordering are relatively simple in polymers consisting of unbranched and stiff chains, of which the most extreme example is polytetrafluoroethylene. Such polymers crystallize easily and rapidly, and consist at low temperature of crystalline regions interspersed with not very extensive disordered regions. The disordering of such a polymer at high temperature is not so very dissimilar to the melting of a crystalline material which is somewhat impure, and has a melting range rather than a sharp melting point.

Conditions are more complicated for polymers of Type C which consist of very flexible molecules. These polymers are apt to show rubber-like behaviour, which means that coiling and uncoiling of the molecules can be appreciably influenced by an applied mechanical stress[6]. Such polymers are analogous to the associated liquids described in Chapter 11; they show a transition to the glassy state, like these liquids, and are glassy rather than crystalline at low temperature.

The relaxation of flexible macro-molecules is a very complicated process, as may be seen from a schematic illustration by Ferry[7]. The three parts of Fig. 2 represent different modes of relaxation of coiled chains. The relaxation of bent and coiled chains may be approached by way of dilute solutions of macro-molecules[17,18] and extended to concentrated solutions, non-cross-linked polymers and finally cross-linked polymers[19]. The work on this subject has been recently reviewed[6,7,8]. The theory is laborious and demands a judicious use of approximations, since the exact derivation of the partition

FIG. 2. Schematic illustration of characteristic modes of coordinated motion of a flexible chain molecule (from Ferry[7]).

functions and free energies of systems of such complexity is not practicable. Rouse[18] uses a model where balls are connected freely by elastic springs. Figure 3 shows a comparison of the mechanical compliance calculated accord-

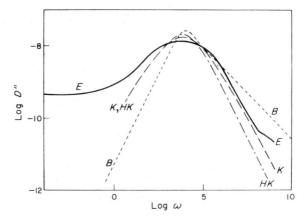

FIG. 3. Comparison of the experimental loss compliance (E) for lightly vulcanized rubber with theoretical predictions (from Ferry[7]).

ing to different models for lightly vulcanized (slightly cross-linked) rubber. It shows that theories agree reasonably well with each other. The data refer to a temperature when the material is in its rubbery high temperature state. The figure shows that for low frequencies the experimental data are not well reproduced by any of the theories.

The glass transition

When an amorphous polymer is cooled through a certain temperature range its properties change to those of a glass. At liquid air temperature rubber can be made to ring like a bell. While for associated liquids the glass transition may be conveniently specified as a change of viscosity to above 10^{13} poises, it may in a rubbery solid be specified as a change of the elastic modulus from some 10^6 dynes cm^{-2} to some 10^{10} dynes cm^{-2}.

The most fundamental description of a glass is thermodynamic[20]. A glass is a material whose entropy does not go to zero at absolute zero temperature. The transition to the glassy state can only be described if the speed of cooling is known. Figure 4 shows the change of structure during the glass transition in terms of changes of density, for different speeds of cooling. In thermodynamic equilibrium, i.e. for infinitely slow cooling, the specific volume would follow the curve A–B_0–C_0 drawn partly as a broken line. This curve

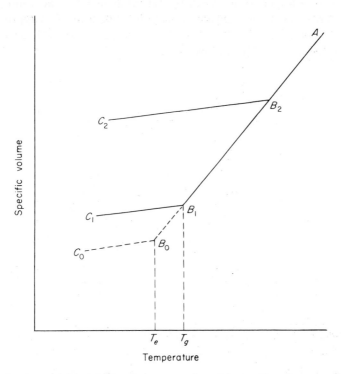

FIG. 4. The glass transition. Schematic representation of the specific volume as a function of temperature for a material in equilibrium (A–B_0–C_0) and in two non-equilibrium glassy forms (modified from Saito[8]).

cannot be straight indefinitely and is here assumed to have a sharp bend which corresponds to some kind of order–disorder transition, in equilibrium, at a temperature T_e. For the two cooling rates 1 and 2 the specific volume does not follow the equilibrium curve, but departs from it at points marked B_1 and B_2 to follow subsequently lines B_1–C_1 and B_2–C_2 which represent properties of a metastable material or glass.

Figure 4 shows that it is difficult to distinguish between the equilibrium and non-equilibrium behaviour of the material. It is usual to call the glass temperature T_g that temperature at which the bend in the measured volume–temperature curve occurs at the slowest practicable rate of cooling exemplified in the figure by A–B_1–C_1. Since the equilibrium curve can, by definition, not be reached in practicable times the existence of an order–disorder transition in equilibrium cannot be directly ascertained.

The glass transition becomes less confusing if we bear in mind the argument in the previous subsection and equation (13.6). If the change in the amorphous polymer is some kind of order–disorder transition, then its relaxation time is likely to increase as the equilibrium transition temperature T_e is approached. The nearness of T_e may be the reason for the sluggishness of the structural changes at temperatures just above T_e.

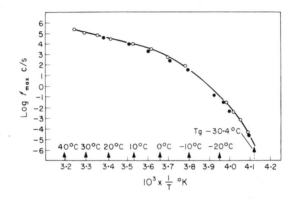

FIG. 5. The frequency of maximum loss in amorphous polyacetylaldehyde as a function of temperature. ○ and ● correspond to two different samples (from Williams[21]).

The difficulty of theoretical treatment of the glass transition lies in specifying just what kind of order–disorder transition is involved. Many theoretical treatments[7,8] use the free volume, i.e. the actual volume less the close packed volume, $v - b$, as in the treatment of the viscosity of liquids, and of rotator solids (see equation 12.1). The results obtained for the behaviour of glycerol under pressure, reported in Chapter 11, indicate that the volume is not the only relevant variable, even in glycerol. However, the free volume is evidently a relevant measure of disorder, and plasticizing liquids act partly by increasing the free volume in a plastic[7].

The relaxation time of the order–disorder phenomenon which is involved in the glass transition may in certain circumstances be measured as a dielectric relaxation time. Williams[21] investigates the amorphous polymer polyacetylaldehyde, where the dipole moment is in the chain backbone. Williams finds relaxation peaks which are not too broad (Cole–Cole parameter α

around 0·6) and uses both a.c. and d.c. techniques. Figure 5 shows the reciprocal of the relaxation time as a function of reciprocal temperature for this polymer. The relaxation time increases in a non-linear way as T_g is approached, as would be expected for an order–disorder transition.

Williams interprets his results in terms of a theoretical treatment by Gibbs and di Marzio[22], which operates with a partition function derived for a flexible linear polymer on the basis of simplifying assumptions. This partition function, which leads to expressions for thermodynamic functions such as the specific heat and the thermal expansion in equilibrium, contains several parameters, for example the energy difference between certain configurations of the chain and the chain length. Gibbs and di Marzio find that their theory gives the right dependence of T_g on the average chain lengths in polystyrene, assuming $T_g = T_e$, which is reasonable in first approximation.

Williams measures ε_s/T and the relaxation time as well as the thermal expansion. He attempts to derive the parameters entering into the Gibbs–di Marzio theory from these data and arrives at a value of T_e which is probably about 5°C below his measured $T_g = -30\cdot4°C$.

Polymers cannot unequivocally be separated into crystalline and amorphous ones. The considerations about the glass transition also apply, more or less, to polymers capable of a high degree of crystalline order. For instance, polyethyleneterephthalate (Terylene) with repeating unit

$$-CH_2-CH_2-CO-\langle\rangle-CO-$$

may be prepared either in a largely (about 60%) crystalline or an essentially amorphous form according to the way in which it is cooled through the temperature range around the glass transition. The dielectric properties of this material have been investigated in great detail[23], particularly in its crystalline form.

Figure 6 shows tan δ for dry crystalline Terylene as a function of temperature and frequency. A high relaxation peak occurs above the glass temperature of about 80°C, and another, smaller peak is found at temperatures between $-60°C$ and $+60°C$ according to frequency. Reddish[23] attributes the high temperature peak to the order–disorder phenomenon to which the glass transition applies. He explains the low temperature peak as due to OH dipoles at chain ends, situated in amorphous regions. Reddish's explanation gives a too large loss per OH group present (see Chapter 15, Section C) but it explains the fact that absorbed water increases the loss angle without changing its frequency dependence. Figure 7 shows $\varepsilon''(\omega)$ at 0°C for amorphous and crystalline Terylene, wet and dry.

In a recent paper Reddish[24] investigates the glass transition in polyvinyl chlorides over a range of frequencies and chemical compositions. He finds that the peak value of $\varepsilon'(T)$ at the transition increases with decreasing frequency of measurement. For polarization over 10^4 sec reversible direct current data (see Chapter 6, Section A) show an $\varepsilon'(T)$ peak of twice the maximum height of the peaks obtained with a.c. This high temperature peak, which depends on chlorination, is interpreted as being due to a cooperative interaction in crystalline regions of long rod dipoles with moments normal to the

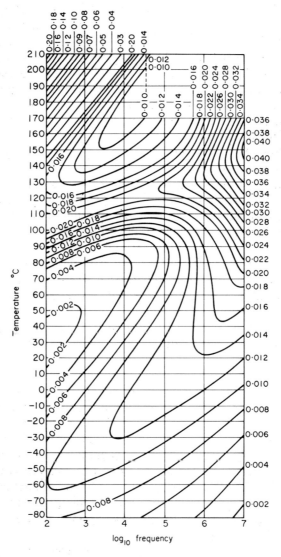

FIG. 6. Tan δ contour map for dry crystalline Terylene (from Reddish[23]).

13. INSULATING MATERIALS WITH AMORPHOUS AND GLASSY CONSTITUENTS 199

chain axes. A low temperature process which is little affected by chlorination involves local movements only.

In spite of the complexities of the glass transition in polymers, $\varepsilon^*(\omega, T)$ may be meaningfully specified for a particular polymer prepared in a particular way as long as T is well below T_g. The loss due to individually mobile dipoles is larger in amorphous than in crystalline regions. The molecular

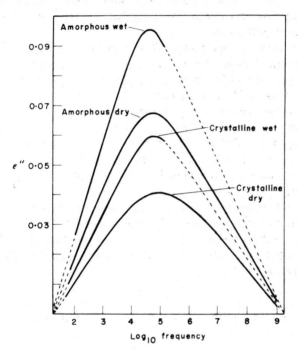

FIG. 7. Plots of ε'' against $\log_{10}$ (frequency) at 0°C for crystalline and amorphous material, both wet and dry (from Reddish[23]).

groups responsible for the dielectric loss may be ascertained by using suitable techniques[23] (for example infra-red spectroscopy). Water may play an important part in the loss, and will be discussed further in Chapter 15, Section C.

B. INORGANIC GLASSES

Inorganic glasses are the amorphous, glassy forms of various silicates, phosphates, etc. The materials[25] are mixed oxides, and form three-dimensional networks. Their bonding is ionic and covalent and the melting points of the crystalline forms are usually high, above 1000°C. Fused SiO_2 which is perhaps the simplest inorganic glass, has a glass transition temperature around 1100°C while glasses which contain Na, K, Ca, Pb, etc. have glass transitions at lower temperatures. The electrical properties of glasses have been surveyed[26] by Stevels, who also surveyed thermal mechanical and other properties[27], and by Sutton[28]. Since inorganic glasses are predominately ionic, they show ionic conductivity and their dielectric properties are of interest only well below the glass temperature. The measurement of dielectric properties becomes complicated when the ionic conductivity is appreciable, as will be discussed in Chapter 14. Nevertheless, glasses may be acceptable dielectrics up to 500°C or so.

The loss mechanisms operating in glasses are in many ways similar to those operating in ionic crystals. The movements of ions into vacant sites within a glass may lead either to d.c. conductivity or to losses of the kind discussed in Chapter 12, Section B, where a vacant site may be filled by an ion in its vicinity, but movements over large distances are topographically or energetically unfavourable. The relevant relaxation time in silicate glasses is governed[26] by equation (4.47) with[26] $A \sim 10^{-13}$ sec and an activation energy of around 17 kcal/mole. This value for silicate glasses is a little higher than the value of around 12 kcal/mole for doped alkali halides as might be expected in view of the stronger cohesive forces.

Apart from the losses connected with the movements of lattice vacancies, glasses also show losses with activation energies of the order 0·1 eV (2·3 kcal/mole). Some of these losses may be of the kind investigated by Stevels and Volger[29] which are illustrated in Fig. 12.10. Other losses of similar activation energy are attributed to small displacements of oxygen atoms, that is deformations of the network[26,28]. Stevels[26] also distinguishes "vibration" losses as due to vibrations of ions in their own interstices. Such movements imply resonance if they are not over-damped. The frequency of vibration loss peaks is the lower the heavier the ions responsible and the more loose the network.

The properties of glasses are greatly affected by water. The fundamental aspect of the behaviour of OH and OD groups in quartz, and migration of protons through quartz have been investigated in great detail by Kats[30]. Some glasses allow very easy diffusion of hydrogen; and adsorbed water causes d.c. conduction along glass surfaces[26,28].

Ceramics, as distinct from glasses, consist of inorganic crystals with glassy admixtures. The properties of ceramics are too various to be considered here.

REFERENCES

1. A. J. Curtis (1960). *Progr. Dielect.*, **2**, 31.
2. R. von Hippel (ed.) (1954). "Dielectric Materials and Applications." John Wiley, New York.
3. R. von Hippel (ed.) (1959). "Molecular Structure and Molecular Engineering." John Wiley, New York.

4. J. B. Birks (ed.) (1960). "Modern Dielectric Materials." Heywood, London.
5. C. G. Garton (1951). *Proc. Instn elect. Engrs*, **98** II, 728.
6. V. Tobolski (1960). "Properties and Structure of Polymers." John Wiley, New York.
7. J. D. Ferry (1961). "Viscoelastic Properties of Polymers." John Wiley, New York.
8. N. Saito, K. Okano, S. Iwayanagi and T. Hydeshima (1963). "Solid State Physics" (ed. Seitz and Turnbull), Vol. 14, p. 343. Academic Press, New York and London.
9. I. T. Barrie, K. A. Buckingham and W. Reddish (1966). *Proc. Instn elect. Engrs*, **113**, 1849.
10. B. Szigeti. Personal communication.
11. D. Sautter (1960). *Prog. Semicond.*, **4**, 125.
12. D. A. Bell (1962). *Proc. Phys. Soc.*, **82**, 117.
13. C. G. Garton (1946). *Trans. Faraday Soc.*, **42**A, 56.
14. H. Pelzer (1957). *J. Polym. Sci.*, **25**, 51.
15. H. Thurn and K. Wolf (1956). *Kolloidzeitschrift*, **148**, 16.
16. H. N. V. Temperley (1956). "Changes of State." Cleaver-Hume Press, London.
17. J. G. Kirkwood and R. M. Fuoss (1941). *J. Chem. Phys.*, **9**, 329.
18. P. E. Rouse (1953). *J. Chem. Phys.*, **21**, 1272.
19. F. Bueche (1954). *J. Chem. Phys.*, **22**, 603.
20. R. O. Davies and G. O. Jones (1953). *Adv. Phys.*, **2**, 370.
21. G. Williams (1963). *Trans. Faraday Soc.*, **59**, 1397.
22. J. H. Gibbs and E. A. di Marzio (1958). *J. Chem. Phys.*, **28**, 373.
23. W. Reddish (1950). *Trans. Faraday Soc.*, **46**, 459.
24. W. Reddish (in press). *J. Polym. Sci.*, Part C.
25. W. Eitel (1956). "The Physical Chemistry of the Silicates." University of Chicago Press, Chicago.
26. J. M. Stevels (1957). "Encyclopedia of Physics" (ed. S. Flügge), Vol. 20, p. 350. Springer-Verlag, Berlin.
27. J. M. Stevels (1962). "Encyclopedia of Physics" (ed. S. Flügge), Vol. 13, p. 510. Springer-Verlag, Berlin.
28. P. M. Sutton (1960). *Progr. Dielect.*, **2**, p. 115.
29. J. M. Stevels and J. Volger (1962). *Philips Res. Rep.*, **17**, 283.
30. A. Kats (1962). *Philips Res. Rep.*, **17**, 133, 201.

CHAPTER 14

Heterogeneous Dielectrics, Non-Ohmic Conduction and Electrode Effects

The present chapter deals with experimental situations where the response of a dielectric to changing fields is complicated by effects which would not occur in a homogeneous material of infinite extension. In many cases these effects are negligible. In others they are important but cannot be treated theoretically because the systems in question contain too many unknown parameters. The cases treated here were chosen from those examples where the theoretical treatment of a complicating effect is relatively straightforward.

A. HETEROGENEOUS MIXTURES WITH LINEAR RESPONSE; THE MAXWELL–WAGNER EFFECT

Dielectrics may consist of conglomerates of macroscopic volume elements with different dielectric constants and conductivities. Such conglomerates may in practice be very complicated since dielectric constants as well as conductivities may be anisotropic, and conductivities need not be linear. In this section it will be assumed that all conductivities are ohmic, and that all phenomena can be described in terms of a macroscopic electrostatic and electrodynamic treatment. With these assumptions it would in principle be possible to derive the dielectric relaxation of any mixture from its material constants and geometry. However, the inverse problem is in general insoluble, and we shall discuss only a few relatively simple examples.

Conducting inclusions in a dielectric may have a very large effect on its electrical response. Resonance effects will occur if the dimensions of the inclusions are of the same order as the wavelength of an alternating field. This is utilized for the construction of artificial dielectrics which may have dielectric constants smaller than unity at microwave frequencies[1].

If the wavelength of the alternating field in question is large compared with the dimensions of conducting bodies and if the conductivity of the inclusions is ohmic, the dielectric response of the heterogeneous dielectric is governed by relaxation equations, i.e. linear first-order differential equations. Relaxation times and polarizabilities can be calculated if the conductivities γ_1, and

γ_2 of the two media, their dielectric constants ε_1 and ε_2 and the geometry are known.

The simplest geometrical case was briefly discussed by Maxwell[2], who considered a plate capacitor filled with n dielectrics of dielectric constants and conductivities $\varepsilon_1, \varepsilon_2, \ldots, \varepsilon_n$ and $\gamma_1, \gamma_2, \ldots, \gamma_n$, arranged in layers parallel to the capacitor plates. Maxwell derives the differential equation for the field between the electrodes as a function of the current through the stratified dielectric. He shows that in general this equation is linear and of the nth order, but reduces to the first order if the ratio ε_r/γ_r is the same for all strata. This means that in general the stratified dielectric, as defined, has a linear response with n relaxation times. Maxwell shows that this model can be used to represent the observed data of dielectric relaxation, but does not discount the possibility that a homogeneous substance may be capable of another kind of delayed polarization.

Wagner[3] develops Maxwell's treatment by calculating the complex dielectric constant of an inhomogeneous dielectric where small spheres of material constants ε_2, γ_2 are dispersed in a medium with constants ε_1, γ_1, the density of the dispersion being small enough to make electrostatic interaction negligible. Wagner is aware that he is calculating the properties of a model, but he is inclined to use Ockham's razor to the effect that a molecular theory is not necessary for the purposes of practical applications. Besides, he shows that his model may represent real conditions in some insulating materials. Following Wagner there was a strong tendency among electrical engineers to consider the explanation of dielectric relaxation in terms of molecular structure as unnecessary.

Sillars[4] develops Wagner's model by considering inclusions which are spheroids, but he simplifies matters by assuming that the conductivity of the medium containing the spheroids is $\gamma_1 = 0$. This calculation is particularly instructive in showing the effects of conductive inclusions. Sillars assumes that the distances between spheroids are large enough to make electrostatic interaction negligible. The inclusions occupy a fraction q of the total volume. The spheroids are assumed to be ellipsoids with axis a in the field direction and equal axes $b = c$ at right angles to the field. In this way the geometry of the heterogeneous dielectric is represented by only two variables, namely the volume fraction q and the axial ratio a/b.

In the static case and for $\gamma_1 = \gamma_2 = 0$ the field inside and outside the spheroid can be derived from field theory as sketched in Chapter 8. The key to this treatment is equation (8.3) which for the present purpose will be written

$$\varepsilon_1 E_1 = \varepsilon_2 E_2 \tag{14.1}$$

where E_1 and E_2 are the normal components of the field at the interface between the dielectrics 1 and 2. For $\gamma_2 \neq 0$ surface charges σ are held at the interface of magnitude given by

$$\sigma = (\varepsilon_1 E_1 - \varepsilon_2 E_2)\varepsilon_0 \qquad (14.2)$$

where σ and E are in practical units, ε_1 and ε_2 in e.s.u. (see Appendix I). The field E_2 within the spheroid is zero in the static case, but not so in the dynamic case, since surface charges are adjusted towards equilibrium by means of currents flowing in the conductive medium. This is the case even if γ_1 as well as γ_2 is finite. For periodic fields when the normal components at the interface may be represented by the complex quantities

$$E_1^*(t) = a^* e^{i\omega t}; \qquad E_2^*(t) = b^* e^{i\omega t} \qquad (14.3)$$

we may calculate the surface charge carried by conduction as

$$\begin{aligned}\sigma^*(t) &= \int (\gamma_1 E_1^* - \gamma_2 E_2^*)\, dt \\ &= (\gamma_1 a^* - \gamma_2 b^*) \int e^{i\omega t}\, dt \\ &= \frac{1}{i\omega}(\gamma_1 E_1^* - \gamma_2 E_2^*) + \text{const.} \end{aligned} \qquad (14.4)$$

where the additive constant may be neglected if the field was switched on a long time before the calculations. If we generalize equation (14.2) by introducing a complex quantity for the charge density, we have another equation for the charge density, namely

$$\sigma^*(t) = (\varepsilon_1' E_1^*(t) - \varepsilon_2' E_2^*(t))\varepsilon_0 \qquad (14.5)$$

where ε_1' and ε_2' are the real parts of the dielectric constants of the two media. Elimination of $\sigma^*(t)$ from equations (14.4) and (14.5) leads to

$$\left(\varepsilon_1'\varepsilon_0 - \frac{i\gamma_1}{\omega}\right) E_1^* = \left(\varepsilon_2'\varepsilon_0 - \frac{i\gamma_2}{\omega}\right) E_2^* \qquad (14.6)$$

which is a generalization of equation (14.1) where the real dielectric constants are replaced by the complex quantities in brackets.

Equation (14.6) may be used to generalize the results derived for fields and charges in the static case to dynamic conditions.

We require here the relationship between the field applied to a capacitor containing the heterogeneous dielectric and between the polarization in the dielectric. The task of deriving this relationship is much simplified if $\gamma_1 = 0$ that is the inclusions are embedded in a perfect dielectric. In this case, all the power dissipation occurs in the inclusions and the complex dielectric constant of the heterogeneous dielectric can be deduced from this power loss. To derive the power loss Sillars uses Ohm's law, so that the power loss per unit volume of the inclusion is

$$W = \tfrac{1}{2}\gamma_2 |E_2^*|^2 \tag{14.7}$$

where E_2^* is the field within the inclusion. For a spheroid as defined this field may be calculated on the basis of field theory, using the generalization expressed in equation (14.6). For a heterogeneous dielectric as defined, containing a volume fraction q of spheroids of a given axial ratio, the calculation leads to a complex dielectric constant of Debye type, namely

$$\varepsilon' = \varepsilon_\infty + \frac{\varepsilon_1' N}{1 + \omega^2 \tau^2} \tag{14.8}$$

$$\varepsilon'' = \frac{\varepsilon_1' N \omega \tau}{1 + \omega^2 \tau^2} \tag{14.9}$$

where ε_∞, N and τ are quantities which involve a dimensionless function $\lambda(a/b)$ of the axial ratio. This function has been evaluated by Sillars; it equals unity for $(a/b) = 0$ and is represented in graphical form in Fig. 1. The expressions for the constants in equations (14.8) and (14.9) are

$$\tau = \frac{(\varepsilon_1'(\lambda - 1) + \varepsilon_2')\varepsilon_0}{\gamma_2} \tag{14.10}$$

$$N = q \frac{\lambda^2 \varepsilon_1'}{\varepsilon_1'(\lambda - 1) + \varepsilon_2'} \tag{14.11}$$

$$\varepsilon_\infty = \varepsilon_1' \left(1 + q \frac{\lambda(\varepsilon_2' - \varepsilon_1')}{\varepsilon_1'(\lambda - 1) + \varepsilon_2'}\right). \tag{14.12}$$

In these equations $\varepsilon_1'(\omega) = \varepsilon_1$ if medium 1 is assumed a perfect dielectric. The frequency dependence of $\varepsilon_2'(\omega)$ raises a complication for reasons outlined in Chapter 6, Section A. The value of $\varepsilon_2'(\omega)$ calculated for low frequencies from the Kramers–Kronig relations may be very large if γ_2 is large and charge carriers pile up at the interface of medium 1 and 2. Effects of this kind are discussed in Sections B and C of this chapter, but ε_2' is normally considered constant.

Sillar's treatment includes the cases of a layered dielectric $\left(\dfrac{a}{b} = 0\right)$ and of spheres ($a = b$) as special cases, although his formulae differ in minor details from those given by Wagner[3] because of slightly different assumptions.

The results for spheroids show a very large influence of the shape of the inclusions on dielectric properties. This implies that little can be deduced from the dielectric properties of a heterogeneous mixture as long as the shape of the conducting particles is not known.

Sillars[4] treatment neglects the electrostatic interaction between conducting particles. The importance of this neglect was investigated by Kharadly and Jackson[5]. These authors consider two and three dimensional arrays of conducting spheres, cylinders, etc. in a loss free dielectric. They give a formula

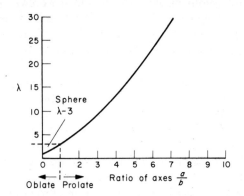

FIG. 1. The dimensionless factor λ in equations (14.10)–(14.12) as a function of the axial ratio $\frac{a}{b}$ of conducting ellipsoids (from Sillars[4]).

for the complex dielectric constant ε^* of a composite dielectric containing a cubic array of spheres which may be written

$$\frac{\varepsilon^*}{\varepsilon_1} = 1 + \frac{\dfrac{3q(\varepsilon_2^* - \varepsilon_1)}{\varepsilon_2^* + 2\varepsilon_1}}{1 - \dfrac{q(\varepsilon_2^* - \varepsilon_1)}{\varepsilon_2^* + 2\varepsilon_1}} \qquad (14.13)$$

where

$$\varepsilon_2^* \varepsilon_0 = \varepsilon_2 \varepsilon_0 - \frac{i\gamma_2}{\omega}.$$

Equation (14.13) is a generalization of equations (14.8) and (14.9), and reduces to the latter for $q \to 0$. Equation (14.13) represents an extension of the Clausius–Mossotti equation, which is valid for a cubic array of point dipoles without spontaneous thermal motion (see Chapter 8). Macroscopic spheres approximate this condition, since a conducting sphere in a uniform field is equivalent to a point dipole pointing in direction of the field. However, equation (14.13) is not exact if the distances between spheres are not large compared with their radii[1].

Jackson and Kharadly verify equation (14.13) experimentally for an artificial dielectric consisting of one layer in which spheres are arranged in a square array. They find that up to $q = 10\%$ the calculation without electrostatic interaction gives moderately good agreement while above $q = 30\%$ even equation (14.13) does not give very good agreement. Dryden and Meakins[6] verify the same theoretical predictions[5] using suspensions of water in woolwax, and suspensions of more or less moist woolwax in petroleum jelly. In the first case where water droplets are fairly good spheres they find fair agreement with equation (14.13) up to about $q = 30\%$. In the second case the agreement with theory was poor. The authors ascribe this to variations in the shape of the conducting inclusions.

Many models of heterogeneous dielectrics have been evaluated; de Loor[7] gives a bibliography which is mainly concerned with the dielectric constant, but he also discusses the relaxation behaviour, in particular the effect of heterogeneity on the distribution of relaxation times. Not all models are useful for the evaluation of practical results; for instance, an appreciable conductivity γ_1 is likely to imply space charges near electrodes and thus non-linear effects, as will be discussed in Section C of this chapter.

The Maxwell–Wagner effect complicates the interpretation of dielectric measurements for materials which undergo a phase change. In particular a not quite pure solid at temperatures not very far below its melting point often consists of solid grains coated by a thin liquid film. Since ionic impurities are likely to be more soluble in the liquid than in the solid, these liquid layers are likely to be relatively conductive. Dielectric constants and dielectric losses may be great near phase transitions for a variety of reasons[8] even if materials are highly purified, and the Maxwell–Wagner effect caused by impurities adds a complication which may be very inconvenient[9]. This applies also to solid–solid transitions, though to a smaller extent than to melting.

B. FIELD DEPENDENT DIELECTRIC LOSS IN THIN LIQUID FILMS

The models described in Section A assume that polarization and field are always linearly related, or else the treatment stops short where non-linearities appear. An inclusion which is a thin liquid film containing ions is a relatively simple case where non-linear effects are important and where a dynamic theory is feasible.

Garton[10] considers a thin layer of a slightly ionized liquid between parallel plane surfaces, with the applied field normal to the layer. The field pulls positive ions to the negative layer surface and *vice versa*. In a static field the ions form more or less separate positive and negative layers, respectively, at the two solid surfaces of the liquid layer unless they can discharge at the surface. The agglomeration of ions causes an electrical field which opposes the applied field. It is assumed that the ions cannot discharge and that the space charges due to ions cause fields which are negligible compared with the applied field.

What happens in an alternating field depends on the distance an ion traverses during a half cycle. Figure 2 shows the force acting on an ion and the path of an ion, as a function of time. For the given layer thickness ionic mobility and amplitude of the electrical field in the layer, the ion spends part

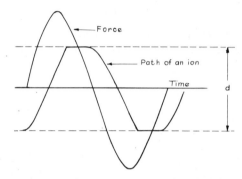

FIG. 2. Force acting on an ion, and paths of ion as a function of time, in a liquid of thickness d, subjected to an alternating voltage (from Garton[10]).

of its time at the surfaces of the layer. This clearly reduces the dielectric loss, since the ion dissipates energy only while it moves under the influence of the field. Hence the dielectric loss, expressed by tan δ, decreases monotonously with increasing field, as long as no other effects come into play.

The saturation effect described above is opposed by another effect as the accumulation of ions of one sign at the surfaces disturbs the dissociation equilibrium. Garton assumes that there are M dissociable molecules present per unit volume and that these molecules dissociate to some extent to form two ions whose charge is equal and opposite. The number of ions present in equilibrium is established dynamically. In the absence of an applied field the number of molecules dissociating or positive ions forming per unit time is given by

$$\frac{dn^+}{dt} = K_D(M - n^+) \qquad (14.14)$$

while the number recombining is

$$\frac{dn^+}{dt} = K_R . n^+ . n^- \qquad (14.15)$$

where K_R and K_D are constants. Equilibrium is established when these numbers are equal and the law of mass action follows for this case as

$$\frac{n^+ . n^-}{(M - n^+)} = \frac{K_D}{K_R}. \quad (14.16)$$

Equation (14.15) is a relaxation equation if n^- may be assumed constant. This is permissible for the simple case assumed by Garton[10] which is discussed in a wider context by Spenke[11]. The relaxation time of the equilibrium is here

$$\tau_{equ} = \frac{1}{K_R n^-}. \quad (14.17)$$

This time is also the lifetime of an ion in the sense of the average expectation of life of all ions which were formed at a given point in time t_0 (see Spenke[11],

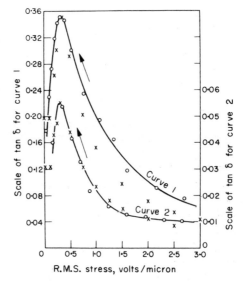

FIG. 3. Relation between tan δ and electric stress. ○ Observed values; × Calculated values.

Section 9.1). Another definition of a lifetime will be used in Chapter 15, Section B.

Garton[10] expresses the influence of an applied field on the ionic contribution to the dielectric loss as a modification of the lifetime in (14.17). He introduces a function g and puts for the average lifetime of an ion in the thin layer

$$\tau = \frac{g}{K_R n^-}. \quad (14.18)$$

He evaluates this function g assuming that $\tau \gg 1/\omega$ and calculates tan δ.

The resulting equation for tan δ as a function of the field is explicit but too complicated to be quoted here. It contains as variables the thickness of the layer, the number and mobility of the ions and the frequency and amplitude of the electrical field. For certain combinations of these variables tan δ as a function of the electrical field goes through a pronounced maximum. Figure 3 shows experimental data compared with theory.

The field dependence of the loss may be used to determine the mobility of the ions in question. The procedure is applicable in the practically important case of impregnated paper capacitors where the liquid is held in pores of the paper. This method is capable of considerable precision[12], in that the mobility of ions determined from the observation of diffusion agrees well with their mobility determined from loss measurements. The phenomena discussed in this section are sometimes described as the Garton effect.

C. ELECTRODE EFFECTS

The non-linear behaviour of dielectric loss in a thin liquid film is a relatively simple example of electrode effects at interfaces. In general, charge carriers in solids or liquids may be electronic as well as ionic and conditions at electrodes may be extremely complicated.

Complications due to electrodes in dielectric measurements are the more troublesome the more conductive the dielectric in question. This is a loose statement in that a dielectric has in theory no direct conductivity at all. In practice, one tends to define a good dielectric as one with $\rho > 10^{15}\ \Omega$ cm. Materials with $10^4 < \rho < 10^{12}\ \Omega$ cm may be classed among dielectrics or semi-conductors, while materials with $\rho < 10^4\ \Omega$ cm are normally considered semi-conductors or metals.

The best understood interface, in the electrical sense, is the junction between two regions of the same semi-conductor, a p–n junction. The n and p regions have different Fermi levels at a large distance from the junction. Near the junction the charge density and potential vary steeply with distance in thermodynamic equilibrium. The theory of p–n junctions is treated in text books on semi-conductors[11].

A p–n junction has a capacitance. For any given voltage applied externally to the interface or junction there exists a certain equilibrium distribution of charge carriers near the interface. If the voltage across the junction is changed, this charge distribution has to readjust itself. This means that during a change of the externally applied voltage the current is not equal to that current which corresponds to the instantaneous voltage, but that there is an additional current proportional to dV/dt. Such a current is a capacitative current and we have therefore to attribute a certain capacitance to the junction[11].

Capacitative currents may be observed not only at p–n junctions in semi-conductors but also other interfaces. In particular at the electrodes applied to an electrolyte, a molten salt or a glass. Some aspects of these effects may be understood in terms of the bulk semi-conductors or electrolytes in question.

The bulk properties of interest here may be stated simply for those cases where Boltzmann statistics applies, that is for ionic conductors and, broadly, non-degenerate semi-conductors. The argument can be based on the idea that diffusion opposes the electrical field, as outlined in Chapter 12, Section B. Equation (12.11) defines stable equilibrium as the condition where the diffusion current just balances the current due to the applied field, so that the net current is zero. When the equilibrium is disturbed a current flows. This current must obey the continuity equation

$$\text{div } I = -e \frac{\partial n}{\partial t} \qquad (14.19a)$$

which may also be written for flow in the x direction

$$\frac{\partial I}{\partial x} = -e \frac{\partial n}{\partial t} \qquad (14.19b)$$

where en is the volume charge density. Besides, Poisson's equation connects field and charge density as

$$\text{div } E = \frac{en}{\varepsilon \varepsilon_0}. \qquad (14.20)$$

When the equilibrium charge density is disturbed, say by injecting a number of charge carriers into the bulk of an earthed semi-conductor crystal, the excess charge cannot persist. A relaxation time may be derived for the decay of the excess charge but the derivation is in general complicated by the presence of different species of carriers. For the simplest case of carrier injection (Spenke[11], Section V.2) equations (14.19) and (14.20) lead to

$$\tau = \varepsilon \varepsilon_0 \rho \qquad (14.21)$$

where ρ is the volume resistivity.

Injection of carriers may introduce a temporary deviation of carrier density from equilibrium. A deviation may also be introduced permanently in space, if, say, $n = n_0$ in the bulk material in equilibrium and if an excess value $n_0 + \mathrm{d}n$ is maintained for $x < 0$. A situation similar to this applies in a simple case of a p–n junction (Speake, Section IV.8) where it can be shown that the time independent value for $n(x)$ is given by

$$n(x) = n_0 + \text{const.} \, e^{-\frac{x}{L}}. \qquad (14.22)$$

The characteristic length is

$$L = \sqrt{D\tau} \tag{14.23}$$

where D is the coefficient of diffusion and τ a relaxation time analogous to that defined by equation (14.21). Characteristic lengths can be defined also for less simple situations, where more than one carrier is mobile. Equation (14.23) indicates the order of magnitude of the length over which a deviation in charge density diminishes.

Equation (14.21) is appropriate for the description of a fluctuation of charge density in a bulk material. This equation is identical with (14.10) if we put $\lambda = 1$, that is assume a flat ohmically conducting layer sandwiched in a dielectric. The dielectric constant in equation (14.21) is that of the material through which the carriers move. As in equation (14.10) there is some ambiguity here, insofar as the presence of carriers may modify the effective dielectric constant between them.

Conditions at an interface are relatively simple in the case of the dilute electrolyte discussed in Section B. In the model treated by Garton[10] no assumption is made about the surface other than that ions cannot discharge. The accumulation of ions near the two surfaces of the layer implies two space charge layers which are assumed[10] to make a negligible contribution to the electric field. The concentration of ions is assumed so low that the characteristic length is large. However, the structure of the interface complicates matters very greatly as soon as L is not very large compared with atomic dimensions. No simple theory of the space charge layer exists even for electrolytes[13] and glasses[14], for conditions where capacitative effects at surfaces are appreciable. Semi-conductors are even more complicated.

All that can be said in general about the space charge and its capacitative effects is that $n(x)$ near the interface varies in a very non-linear way, and that $n(x)$ depends on material constants of the two adjoining materials as well as on the applied field. However, it is possible to define a total space charge Q by some integral of $n(x)$ over x. The voltage dependence of Q defines an incremental capacitance[11]

$$C = \left(\frac{\partial Q}{\partial V}\right)_{V=V_0}. \tag{14.24}$$

This capacitance depends on the direction as well as the magnitude of V_0.

Bardeen[15] gives a lucid discussion of the a.c. impedance of a contact rectifier, that is a contact between a metal and a semi-conductor. One relaxation time for his model is given by equation (14.21). This time is only of the order 10^{-10} sec for a semi-conductor with $\rho = 100\ \Omega$ cm. However, Bardeen points out that much longer times may be required for readjustments of donor and acceptor ions.

Barrier layer capacitances may be very large[13,14]. A good example is rutile (TiO$_2$) which can be made conductive by removal of oxygen. While $\varepsilon_s = 170$ or so for an insulating crystal in the crystallographic c direction much larger "apparent" dielectric constants are measured for conducting crystals. Parker[16] investigates the dielectric behaviour of rutile and finds, in some cases, a fair approximation to Debye peaks, with, for instance, $\varepsilon_s = 3 \times 10^4$. The position of the relaxation peak corresponds tolerably well to equation (14.21). This behaviour is independent of the electrode material, and the large measured ε_s and its relaxation can be attributed to the space charge in rutile. However, a straightforward result of this kind is unusual with contact effects.

In good, or even moderately good, dielectrics the role of electrode effects is relatively minor, but need not be negligible. Any dielectric should strictly speaking be represented by the circuit in Fig. 4, where arrows represent voltage dependence.

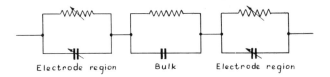

FIG. 4. Equivalent circuit of a dielectric with electrodes which are not ohmic.

The role of the electrodes for a fairly good dielectric was examined experimentally by Ben Sira and co-workers[17]. They used a long slab of coloured, photoconductive NaCl, where illumination of a strip in the centre was equivalent to the connection of a low resistance in parallel with the illuminated region. The authors applied a direct voltage to the electrodes and observed the changes of current after the illumination of a strip was switched on or off. The measurements were used to calculate the drift length $\mu\tau$ of the carriers (μ being their mobility and τ their lifetime). Incidentally, the authors also determined the field along the sample due to the application of an external voltage, as shown in Fig. 5. The figure shows that the field is highest at the electrodes and that the field distribution depends on the material of the electrodes. In the limiting case where the field is constant against distance the electrodes would be non-blocking or ohmic.

In another recent investigation of NaCl single crystals the authors[18] came to the conclusion that electrode effects are unimportant with regard to dielectric measurements. These authors measured the response of electroded specimens to a step-function in voltage, and found the measured decay linear with voltage. They also coloured a layer of the crystal near the surface with

X-rays, thereby making the two electrode regions different, and found no dependence of the discharge current on the direction of the field.

The discrepancy between the conclusions of the two investigations quoted is not surprising. In the case of a good dielectric with a d.c. conductivity of under 10^{-15} Ω^{-1} cm^{-1} the impedance of the sample will be governed by the resistance of the central, bulk region (see Fig. 4). The electrode regions become important only if the resistance of the central region is relatively low. Jacobs and Maycock[19] extend measurements on alkali halide crystals to temperatures around 600°C, where resistivities are lower, dope their crystals with divalent impurities, and use specimens of different thickness. They find voltage dependent capacitances of contact layers.

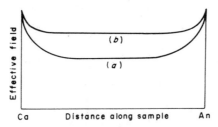

FIG. 5. The electric field distribution along a sample of NaCl due to the application of an external voltage. (a) with silver paint electrodes; (b) with indium electrodes (from Ben Sira et al.[17]).

Jacobs and Maycock determine the capacitance of the dielectric between two electrodes as a function of the specimen thickness, the conductivity and the frequency. In this context they measure low field capacitances, i.e. they use fields small enough to permit the neglect of voltage dependence. The results give satisfactory agreement with theory.

Barrier layers at electrodes become important in a practical context at high temperatures where all dielectrics become conductive and conductivities of the order $\gamma = 10^{-6} - 10^{-4}$ Ω^{-1} cm^{-1} have to be tolerated in electrical insulation. In this case the dielectric loss as well as the practically useful life of insulation may depend on the electrode material[20].

Conditions are most complicated for granular semi-conducting materials, such as non-stoichiometric oxides. Here barrier layers may arise at the interfaces between grains, and the dielectric loss of the bulk material may be markedly dependent on a biasing d.c. field. Volger[21] gives a survey of this type of dielectric with reference to low field relaxation and shows that a two layer model gives a simple and useful representation. He also considers the field dependence of relaxation and more complicated effects such as the effect of light on intergranular barrier layers. Kröger[22] (Section 22.11) gives a general discussion of polarization and capacitance effects at electrodes with a bibliography up to 1963.

REFERENCES

1. J. Brown (1960). *Prog. Dielect.*, **2**, 193.
2. J. C. Maxwell (1954). "A Treatise on Electricity and Magnetism", Vol. 1, p. 452. Dover Publications, Dover.
3. K. W. Wagner: (1914). *Arch. Elektrotech.*, **2**, 371.
4. R. W. Sillars (1937). *J. Proc. Instn. elect. Engrs*, **80**, 378.
5. M. M. Z. Kharadly and W. Jackson (1953). *Proc. Instn. elect. Engrs*, **100** III, 199.
6. J. S. Dryden and R. J. Meakins (1957). *Proc. Phys. Soc.*, B, **70**, 427.
7. G. P. de Loor (1956). "Dielectric Properties of Heterogeneous Mixtures." Thesis, Leiden.
8. H. N. V. Temperley (1956). "Changes of State." Cleaver-Hume Press, London.
9. V. Daniel (1952). E.R.A. Report L/T281.
10. C. G. Garton (1939). E.R.A. Report L/T108 (1941). *Proc. Instn. elect. Engrs*, **88**, III, 23.
11. E. Spenke (1958). "Electronic Semiconductors." McGraw Hill, New York.
12. Z. Krasucki (1965). E.R.A. Report, Ref. No. 5070.
13. P. Delahay (1965). "Double Layer and Electrode Kinetics." Interscience, New York.
14. J. Volger, J. M. Stevels and C. van Amerongen (1953). *Philips Res. Rep.*, **8**, 452.
15. J. Bardeen (1949). *Bell Syst. tech. J.*, **28**, 428.
16. R. A. Parker and J. H. Wasilik (1960). *Phys. Rev.*, **120**, 1631.
17. M. Y. Ben Sira, B. Pratt, E. Harnik and A. Many (1959). *Phys. Rev.*, **115**, 554.
18. P. H. Sutter and A. S. Nowick (1963). *J. appl. Phys.*, **34**, 734.
19. P. W. M. Jacobs and J. N. Maycock (1963). *J. Chem. Phys.*, **39**, 757.
20. V. Daniel and M. G. Rogers (1964). "Special Ceramics" (ed. P. Popper), p. 187. Academic Press, New York and London.
21. J. Volger (1960). *Prog. Semicond.*, **4**, 205.
22. F. A. Kröger (1964). "The Chemistry of Imperfect Crystals." North Holland Publications, Amsterdam.

CHAPTER 15

Hydrogen Bonded Solids

The hydrogen bond has already been defined and discussed in Chapter 11 in connection with water and associated liquids. Its role in solids is no less important. To recapitulate, the hydrogen bond may be symbolized as R_1—H---R_2, where R_1 and R_2 are electronegative atoms. While the full line R_1—H signifies a homopolar bond, the dotted line R_2---H symbolizes an anomalous bond caused by a combination of dipolar attraction and quantum mechanical exchange forces.

The hydrogen bond plays an important part in the electrical properties of matter in general because of its variability. This variability is obvious in a liquid where hydrogen bonds are constantly broken and re-formed. However, even in a solid hydrogen bonds may occur in polymeric forms, chains, rings and networks, and the difference of energy between different configurations may be so small that the configurations fluctuate spontaneously even at or below room temperature. In cases where different configurations differ in their polarization the dielectric constant may be high and an electric field may influence the molecular structure to an unusually large extent.

Some hydrogen bonded solids are ferroelectric and these will be discussed in the next chapter. Here we shall treat two relatively well understood cases of unusual dielectric properties in materials which are not ferroelectric, and consider briefly the very important practical role played by moisture in the behaviour of solid insulation.

A. SECONDARY LONG-CHAIN ALCOHOLS

A typical compound of this group has the chemical formula $CH_3(CH_2)_p CHOH(CH_2)_q CH_3$ where p and q are integers. The alcohols crystallize in variants of the basic paraffin structure (see Chapter 12) so that OH groups lie on planes approximately normal to the paraffin molecules (see Fig. 1). Within such a plane hydrogen bonding (indicated by dotted lines) occurs between neighbouring OH groups. The energy contribution from this bonding causes the melting points of the alcohols to be appreciably higher than those of the corresponding paraffins.

The curious dielectric properties of this group of materials were discovered by Meakins and co-workers. The characteristic feature is an anomalously high dielectric constant which decreases with time[1] and which is anisotropic[2], being larger in the plane of the OH groups than normal to it. The anomalously

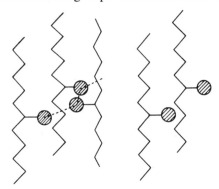

FIG. 1. Schematic structure of a secondary long-chain alcohol (from E.R.A. House Journal, April, 1962). ⟨ Paraffin molecule; ● Oxygen; ··· Hydrogen bonding.

high dielectric constant depends on the way in which a given sample has been prepared, as may be seen in Fig. 2. The figure shows that material recrystallized from solution has a much lower dielectric constant than material cast from the melt, but that the peak of $\varepsilon''(\omega)$ lies in the same region in both cases.

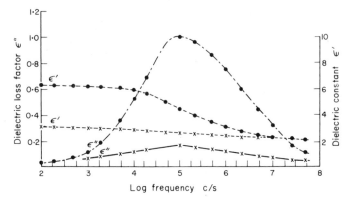

FIG. 2. Dielectric absorption of melted and recrystallized forms of the alcohol $C_{13}H_{27}CHOH$ $C_{13}H_{27}$ (from Meakins and Sack[1]). — Melted form; —×— Recrystallized from ethanol.

The magnitude of ε_s for the form solidified from the melt is such that it implies a dipole moment appreciably higher than the moment μ_0 of the OH group in solution (about 1.6×10^{-18} e.s.u.). The value of the moment per OH group in the solid varies according to the method of calculation used and the material. Values of more than 200 μ_0 have been proposed[3] while other calculations[4] give typical values of $\mu = 4.5$–6×10^{-18} e.s.u. for the average

OH group. In any case the values are too high to be accounted for by normal dipole reversal. The possibility of Maxwell–Wagner effects is ruled out by the care taken by the experimenters and the observation that purification increases rather than decreases ε_s.

Meakins and Sack[1] explain the high values of ε observed on the basis of chains of hydrogen bonds which may reverse their polarity because they contain a discontinuity. This model may be represented as sketched

```
      C                  C                      C
      |                  |                      |
  H—O---H—O          O—H---O—H---O—H
          |              |
          C              C
```

where the plane of the drawing is the plane in which the OH groups approximately lie. The discontinuity in the centre migrates by one link if one of the neighbouring OH groups rotates about the C—O bond, an energetically easy process. The chain can reverse its dipole moment by successive steps. Sack[5,6] gives a quantitative theory of the complex dielectric constant on the basis of the model. He finds[6] for a dielectric containing N dipoles/cm^{-3}, arranged in chains with $n - 1$ dipoles each

$$\varepsilon_s - \varepsilon_\infty = C \frac{\mu^2}{3kT} N . n \tag{15.1}$$

where μ is the dipole moment of a single dipole and C is a constant. A later calculation by a similar method[6] confirms the formula (except that n is replaced by $n + 1$) but while Sack gives $C = 128/\pi^3$ the later calculation gives the slightly different value $C = 4\pi$. Both calculations neglect electrostatic interaction. Sack [5,6] derives the dielectric relaxation of the model and finds a series of relaxation times. For n large Sack finds for the longest relaxation time

$$\tau_n = \frac{2n^2 \tau_0}{\pi^2} \tag{15.2}$$

where τ_0 is the relaxation time for an isolated dipole. The shorter relaxation times have values $\frac{1}{9}$, $\frac{1}{25}$ etc. of τ_n.

Sack's theory does not by itself account for all the experimental evidence. The crystal structure shown in Fig. 1 implies long chains of hydrogen bonds since chains end only at the edges of crystallites. If this structure were correct and all chains of hydrogen bonds were reversible the static dielectric constant would be much higher than observed. Besides, there would be no reason for its decay with time. One must assume that chains either end within crystallites and/or that not all chains are reversible.

Daniel[8,9] proposes an explanation of the data which is based on the fact that unsymmetrical secondary alcohols have optical isomers and that there are many possible arrangements of left and right handed isomers in a chain sequence (symmetrical secondary alcohols behave similarly since chains can be "upside down"). Hence the hydrogen bonding within a plane of OH groups may be disordered in an intricate way; some OH groups are in ambiguous positions, so that they may join either one or another chain. This texture is able to explain all observed data.

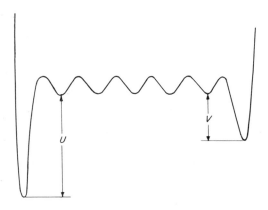

FIG. 3. Potential energy as a function of distance for a chain of hydrogen bonds where the structural configuration favours polarization to the left (from Daniel and Vein[7] report No. 5014 of Electrical Research Association).

The texture proposed may be represented by the potential well model[7] shown in Fig. 3, where the wells at the two ends are deeper than those in the centre and where one end well may be deeper than the other. The depths of these wells may be thought to represent the potential energies associated with OH groups at free ends of chains, in rings at chain ends, and inside chains. The interest of the model represented by Fig. 3 lies in its response to high electrical fields. This response is strikingly different from that of a simple long dipole if the depths of the end wells differ greatly, for instance if the well at the right is shallow with $V = 0$ while the left well characterized by $U = W/kT$ is deep. In this case a chain with a not very large number $m = n - 1$ of dipoles is at low temperatures mostly in a state where all its dipoles point to the left. Only a fraction

$$G = \frac{m}{m + e^{\frac{W}{kT}}} \qquad (15.3)$$

of all chains contribute to the dielectric constant, if this is measured using a negligibly small electrical field. The influence of the chain length on the value of G may be understood as due to the entropy of rearrangements of dipoles within the chain. For an appreciable electrical field acting to the right the entropy in question aids the field and G increases, so that the dielectric constant increases with field strength. For certain parameters m and W the dielectric constant goes through a maximum whose value may be more than three times the low field value[7]. Experiments[9] confirm some of the predictions of this theory, but are not conclusive because of the difficulty of reproducing a given texture.

At this stage it would be desirable to determine the average length m of chains of hydrogen bonds in a secondary alcohol by as many methods as possible. The most reliable method seems to be that of Meakins[3] who measured τ_0 for a dilute solution of the alcohol in question in a paraffin of the same molecular length where the alcohol would form a monomer or dimer. Meakins derives $n = m + 1$ from Sack's formula (15.2). However, at this point there is need for further theoretical work since Sack's formula refers to the relaxation of a chain of dipoles such as may be represented by Fig. 3 if the end wells are equal to the centre wells. Equation (15.2) may need to be modified for the case of deep end wells. Apart from Meakins' method[3] and dielectric saturation, infra-red spectroscopy[4] and paramagnetic resonance[10] also provide information about the length of hydrogen bonded chains.

Sack's theory seems to apply more closely[3] to alicyclic secondary alcohols such as cyclohexanol than to long chain secondary alcohols. The alicyclic compounds show a narrow distribution of relaxation times, the length of hydrogen bond chains is short (3-4 links) and end wells appear equal in energy.

Many biological materials, such as proteins, contain extensive hydrogen bonding, and the long-chain alcohols are a relatively tractable example of this class of material as the hydrogen bonding is in a planar configuration sandwiched within a water repellent paraffin. Chains of hydrogen bonds could be of biological interest because they are able to concentrate the energy of an electrical field stored within a large volume on to a small volume at the end of a chain[7]. A speculative suggestion has been made[11] that an effect of this kind might play a part in nerve membranes.

B. ICE

Ice is a solid bonded exclusively by hydrogen bonds. There are several structural modifications, the most important being the hexagonal one stable near the melting point[12]. In all cases oxygen atoms are at the centres of tetrahedrons, as is the case with carbon atoms in diamond. Figure 4 gives a three dimensional representation which shows that O—H---O linkages hold the tetrahedral network together.

The structure shown is ideal, in that a crystal of indefinite magnitude could be built in the way indicated so that each hydrogen would have one, uniquely determined, position. This bonding obeys the Bernal–Fowler rules[14] which signify that each oxygen has two hydrogens normally bonded to it and two further ones bonded by hydrogen bonds. However, real ice is not as regular

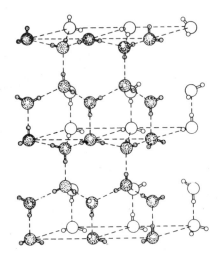

FIG. 4. The arrangement of molecules in hexagonal ice. The orientation of water molecules, as represented in the drawing, is arbitrary. There is one proton along each oxygen–oxygen axis, closer to one or the other of the two oxygen atoms (from L. Pauling[13]).

as this. There are many possible ways of permutating O—H---O configurations in three dimensions so that the positions of hydrogens are not uniquely fixed, and without including more than a very few wrongly bonded oxygens. The entropy of ice at 0°K can be measured by finding the entropy of water vapour in two different ways. If the hydrogens were positioned ideally the zero point entropy would vanish. In fact it is

$$S(0) = 0.805 \text{ cal}/^\circ\text{C} \simeq R \log (\tfrac{3}{2}) \tag{15.4}$$

which indicates that hydrogens are disordered. The magnitude of $S(0)$ has been explained by Pauling, and the argument is summarized by Onsager and Dupuis[15] in a recent survey of the kinetic properties of ice.

Pure ice has at 0°C a dielectric constant of $\varepsilon_s = 105$ in the direction of the hexagonal c axis and $\varepsilon_s = 91$ in the direction normal to it. This high value implies disorder, fluctuations in the directions of O—H bonds. The magnitude of ε_s has been derived by Gränicher[17] and co-workers on the basis of a detailed consideration of the geometry and is also discussed by Onsager and Dupuis[16].

The relaxation behaviour of pure ice has been investigated for polycrystalline material by Auty and Cole[18] who found a single relaxation time. Single crystals were investigated by Humbel[19] *et al.* with the result that the relaxation time was single and independent of direction. At $-10.8°C$ the values for the (extrinsic) relaxation time were $\tau = 5.8 \times 10^{-5}$ sec in the c direction and 6.1×10^{-5} sec normal to it. Equation (4.47) applies; $A = 5.3 \times 10^{-16}$ sec, $W = 13.25$ kcal/mole are recognized values[16] for H_2O, and $A = 7.7 \times 10^{-16}$ sec, $W = 13.4$ kcal/mole for D_2O. The pre-exponential factor A is unusually low. The mechanical relaxation data have been determined[20] for single crystals and agree with the extrinsic dielectric data within the limits of experimental error.

The relationship between the relaxation times of H_2O and D_2O is interesting, in that τ_d for heavy ice is longer by a factor of about $\sqrt{2}$. This ratio is approached in the liquid near $0°C$ (see Chapter 11, Table IV). It may be expected that if O—H and O—D then stretching vibrations play an important part since the frequencies of these vibrations in the two compounds are related as $\sqrt{(2)}:1$.

The dielectric and structural properties of ice pose a fascinating problem. It is clear that the configuration of hydrogens must fluctuate, and that it must do so in a random manner since the relaxation time is single (see Chapter 4). The relaxation time is independent of crystal size, and the re-orientation of dipoles can therefore not involve surfaces. Hence it must be due to the migration of internal defect. What are these defects, and how can they act in a random manner?

A great deal of work has been done on the defects in ice. The relevant electrical measurements concern conductivity as well as dielectric data and the theory as well as the experimentation is exacting and intricate. Dielectric techniques were used in a long series of experiments in Zürich and the results of these have recently been summarized by Gränicher[21,22]. Electrochemical techniques were used in Göttingen and have recently been summarized by Eigen and de Maeyer[23,24]. Both reviews are based on a great deal of original work by the two groups mentioned. Apart from the work summarized in these reviews and the other papers quoted ice has been the subject of a literature too vast to be discussed in detail here. The present monograph aims to give a physical picture of the unusual kind of relaxation exhibited by ice, and does not claim to cover all aspects in full.

As has originally been shown by Bjerrum[25], a single crystal of ice, disordered in the sense discussed above, contains four kinds of defects. Two ionic ones

$$OH_3^+ \text{ and } OH^-$$

and two orientational or Bjerrum defects

$$O\text{—}H \quad H\text{—}O \quad \text{and} \quad O \quad O.$$

The first of the Bjerrum defects is usually called D defect and the second one L defect. It will be seen that the L defect is identical with the defect envisaged in chains of hydrogen bonds in solid alcohols. In the case of such alcohols a D defect seems energetically unlikely[5] because of the mutual repulsion of the hydrogens. In ice L defects can only occur if accompanied by D defects. The latter cause only a relatively small energy of deformation because the three-dimensional network is able to distort so as to take up the strain. Figure 5 shows the two kinds of defects.

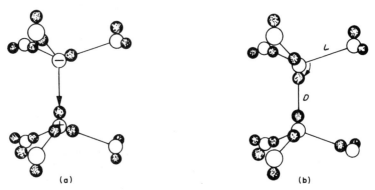

FIG. 5. (a) Formation of a pair of ion states by the translation of a proton along the hydrogen bond; (b) Formation of a pair of L–D defects by a rotational motion of the proton within the same molecule (from Gränicher[22]).

The kinetic behaviour of ice may be explained on the basis of the four types of defects mentioned. In this way, one might say that relaxation in ice is a defect property. However, defect properties are usually erratic because they are due to defects which are not in thermodynamic equilibrium. The kinetic properties of ice are reproducible because they relate to defects which are present in equilibrium.

In thermodynamic equilibrium, the number of ionic as well as of orientational defects is governed by the law of mass action. For orientational defects[23,24] one may write a chemical equation

$$2N \rightleftharpoons D + L \qquad (15.5)$$

and the law of mass action is then expressed as

$$\frac{N_D N_L}{N^2} = \text{const.} \; e^{-\frac{W_0}{kT}} \qquad (15.6)$$

where N_D, N_L and N are the numbers of the defective respectively normal bonds and W_0 is the energy needed to form an L–D pair. For pure ice the number of L and D defects is necessarily equal. A similar law of mass action applies to the ionic defects.

It has been proved[21] that the measured intrinsic conductivity of purest ice is predominantly ionic rather than electronic. This direct conductivity can be explained as due to the movement of the two ionic defects. According to the usual treatment of ionic conductivity each kind of defect may be credited with a mobility of the form

$$\mu = B\,e^{-\frac{W}{kT}} \tag{15.7}$$

where B is a constant and W is the activation energy of the movement of an ion from one site to a neighbouring one. The low field direct conductivity is given in terms of the products of the number and mobility for each ion (see equation 12.14).

The orientational defects are associated with electrical charges in that a D defect implies an excess of positive charge, an L defect an excess of negative charge. These defects cannot sustain a direct current since they cannot discharge at electrodes, but their movement is able to cause a change in the electrical polarization. Gränicher[17,21,22] and co-workers explain the high dielectric constant of pure ice as due mainly to the movements of L–D defects. These faults move by a kind of rotation about one of the oxygens involved in the defective bond. Gränicher defines mobilities of the form (15.7) for the L and D defects.

The relaxation time and low field conductivity for pure ice do not provide sufficient data for the determination of the six activation energies in question —namely the two activation energies of formation of the two defect pairs, and the four activation energies of the four mobilities. Gränicher and co-workers supply additional information by incorporating impurities in ice. Hydrofluoric acid is particularly suitable because fluorine can replace oxygen in ice without altering the structure. Since fluorine is univalent while oxygen is divalent it disturbs the equilibrium of defects in a controlled way.

Eigen and co-workers[23,24] supplement the low field conductivity data by sophisticated electrochemical methods. In particular they apply a high field of short duration to a thin crystal. For sufficiently high fields the current saturates against field strength when ions are swept to the electrodes in a time short compared with that needed to form new ones by dissociation.

The experimental and theoretical work of the authors quoted has so far led to the determination of a number of relevant parameters for the defects. These are summarized in Table I, quoted from Gränicher[22].

It is interesting to consider the kinetic properties of ice from the viewpoint of fluctuations. In general the dielectric relaxation time τ_d is the regression time of the fluctuation in the direction of dipoles (see Chapter 4). However, in ice the dipoles, which signify OH groups, are fixed in position unless their direction is changed as a consequence of the movement of a defect. Bjerrum defects are much more common than ionic defects, and thus the changes in dipole orientation are mostly due to the movements of L and D defects. The movements of defects are very intricate since place changes occur in a three-dimensional network.

The kinetic properties of ice have been discussed by Onsager and Dupuis[15,16] from the viewpoint of dissociation equilibria and their relaxation. These authors also take the three-dimensional character of the problem into account. Onsager and Dupuis base their treatment on the theory of electrolytes, which is touched upon in Chapter 12, Section B and Chapter 14, Section C. Dissociation equilibria of the kind $AB \rightleftharpoons A + B$ also occur in

TABLE I
Pure ice: derived data for $-10°C$

	Bjerrum defects	Ion defects
Reaction equation	$2N \rightleftharpoons D + L$	$2H_2O \rightleftharpoons H_3O + OH^-$
Energy of formation	0.68 ± 0.04 eV	1.2 ± 0.1 eV
		0.96 ± 0.13 eV
Concentration of defects	7×10^{15} cm^{-3}	8×10^{10} cm^{-3}
Activation energy of diffusion	0.235 ± 0.01 eV	~ 0
Transit frequency	2×10^{11} sec^{-1}	6×10^{13} sec^{-1}
Mobility (in cm^2/V sec^{-1})	$\mu_L = 2 \times 10^{-4}$	$\mu^+ = 7.5 \times 10^{-2}$
Mobility ratio	$\mu_L/\mu_D \geq 1$	$\mu^+/\mu^- = 10$–100
Effective charge	$0.6e$	$2.0e$

electronic semi-conductors, when an uncharged atom dissociates into two charged carriers. The relaxation of such equilibria has been discussed by Spenke[26], with particular reference to the meaning of a lifetime for a carrier or ion.

In the present monograph we shall attempt to give a picture of the dynamic properties of ice which may be easily visualized. We shall assume that only Bjerrum defects are important for dielectric relaxation, and that the ice is pure so that L and D defects are present in equal number and in small concentration. The case assumed is analogous to that treated in Chapter 14, Section B. We assume a time independent rate of dissociation as in equation (14.14), with $M \gg n^+$, and a rate of recombination as in equation (14.15). This defines an equilibrium as in (14.16), which is not identical with (15.6). However the difference is unimportant since N is a constant in the given context. Equation (14.17) defines the relaxation time of the equilibrium and we define here a life time $\tau_l = 2\tau_{\text{equ}}$ which is the average time which elapses between the "birth" and "death" of a defect. (This definition is not the same as for the life time in Chapter 14, Section B. The distinction is explained by Spenke[26], Section 9.1.)

In order to simplify the understanding of dielectric relaxation we shall

15. HYDROGEN BONDED SOLIDS

make two further assumptions. We shall neglect the electrostatic attraction or repulsion of the defects, and we shall neglect the geometrical complexities which are present in ice. In the model system each dipole has two opposite directions, and it reverses its orientation at random, in the average once in a time $\frac{1}{2}\tau_d$ where τ_d is the dielectric relaxation time. A change of direction is caused by the passage of a defect across the bond, and hence every bond must be visited by a defect once within a time $\frac{1}{2}\tau_d$, in the average. The relationship between τ_d and the lifetime of the defects is simple if we neglect electrostatic interaction. In that case defects move in the same way whether they are near another defect or not. If we fix our attention on a given bond, we shall in the average see two defects, L or D, arrive at this bond once within a time τ_d. If the bond happens to be defective, say L, the arrival of a D defect will end its life. Now, the definition of lifetime τ_l is that an existing defect "lives" in the average for a time τ_l. Hence $\tau_d = \tau_l$, since we have assumed that a D defect will approach the bond in question in the same way whether the bond is L (that is negatively charged) or not. In the presence of electrostatic interaction τ_d is no longer necessarily equal to τ_l. Onsager and Dupuis[16] take the electrostatic interaction into account and assume that the dielectric constant between two defects is ε_∞, since no relaxation occurs in the intervening matter. They find

$$2\tau_{equ} = \tau_d \frac{\varepsilon_\infty}{\varepsilon_s}. \tag{15.8}$$

Onsager and Dupuis[16] further deduce the dielectric τ_d on the basis of the geometry of ice, using some simplifications. They define v_L and v_D as the frequencies of jumps of L respectively D defects from one bond to the next and find

$$\frac{1}{\tau_d} = \frac{4}{3N}(v_D N_D + v_L N_L) \tag{15.9}$$

where N_L, N_D and N are the number of L defects, D defects and perfect bonds respectively.

The significance of equation (15.9) may be understood in terms of fluctuations with the help of the simplified model. For this model we know that each bond is imperfect only for a fraction $\frac{N_L}{2N} = \frac{N_D}{2N}$ of its existence. We also know that each bond reverses once in a time $\frac{1}{2}\tau_d$. Hence a pair of defects has to visit $\frac{2N}{N_L}$ bonds within that time, each defect half that number. The defect must traverse the bonds it visits so that each transit from a bond to a neighbouring bond takes on average only a fraction $\frac{N_L}{N}$ of the time $\frac{1}{2}\tau_d$. This defines a transit frequency per unit time for the progress of a defect from bond to bond as

$$v = \frac{N}{2N_L} \cdot \frac{1}{\tau_d}. \tag{15.10}$$

Comparison of that equation with equation (15.9) for $N_D = N_L$ shows that the model represents Onsager and Dupuis' relationship between relaxation time and transit frequency in ice quite well. The comparison yields $2v = \frac{4}{3}(v_L + v_D)$. This close relationship between the model and the three-dimensional system does of course not mean that the model represents ice in its microscopic behaviour. The movement of Bjerrum defects in ice proceeds by very complicated three-dimensional paths[20], and it is also influenced by the ionic defects. Besides, the Onsager and Dupuis' equation (15.9) is itself approximate.

Defects in ice move in a random manner along a convoluted path. The speed of defects is in the average given by va where v is the transit frequency of the defect in question and a the length of an O—H---O bond. Using $a = 2.76 \times 10^{-8}$ cm and the data in Table I for transit frequencies at $-10°C$ the speed of Bjerrum defects emerges as of the order 10^4 cm/sec, that of ionic defects of the order 10^6 cm/sec, the order of the speed of sound.

Whalley and co-workers[27] find that the dielectric relaxation time and direct conductivity of hexagonal ice both increase with pressure. This confirms the argument that dielectric relaxation occurs by the movement of orientational defects since such defects should be formed with an increase of volume. The volume of activation for relaxation is found positive and that for conduction negative, in agreement also with Eigen's theory which implies that proton transfer should be facilitated by compression.

Problems still remain with regard to the kinetic properties of ice. In particular, the very high mobility of the positive ionic defects is still under discussion. Eigen and co-workers[23,24] explain this high value as a consequence of quantum mechanical tunnelling. This conclusion is confirmed by the difference they find[24] between water and heavy water, namely $\sim 1 \times 10^{-2}$ cm^2/V sec for D$_2$O against $\sim 8 \times 10^{-2}$ cm^2/V sec for H$_2$O. The theory of transfer between two potential wells by tunnelling, as distinct from a classical hopping process, has recently been discussed by Sussmann[28], with reference to ice and the influence of phonons. However, this is only a first approximation, not taking into account the unique three dimensional bonding of ice.

C. THE ROLE OF HYDROGEN BONDING IN PRACTICAL INSULATION

The most important feature of materials capable of forming hydrogen bonds, by themselves or with water, is the risk of ionic conduction. Ionic conduction may lead to chemical changes either by an accumulation of an ionic species near some compound which can react with it, or by the acceleration of ions to speeds which allow the ion to transfer an energy of more than a few electron volts to some molecule in its way. In either case, permanent chemical damage may result and in practice effects associated with moisture are among the most common causes of the failure of insulation.

It is well known that chemical groups of the form ROH may dissociate into RO^- and H^+ and the dissociation of acids in aqueous solutions is commonplace. Dissociation and conduction by H^+ is less well established in solids but it is clear on chemical grounds that ROH groups in solids have some ionic character, depending not only on the atom to which OH is attached but on the other atoms in the vicinity[13,29]. Insulating materials may contain chemical groups which can give an acid reaction either by themselves or in the presence of sorbed water.

The likelihood of ionic conduction in insulating materials which are free to sorb water depends on their chemical structure. Three types may be distinguished: firstly, water repellent materials which contain no electronegative atoms and which form no hydrogen bonds to water; polyethylene belongs to this class. Secondly, materials which offer sites of attachment for water, but are not themselves hydrogen bonded; polymers with polar groups (see Table I) and ceramics belong to this class. The attachment of sorbed water may be very strong, and ceramics may retain sorbed water up to temperatures of the order 300°C or more. The thermodynamics of sorption has been investigated by Day[30] for polymers used as practical insulation. Polymers containing only a low concentration of polar groups sorb water in the form of clusters of a few molecules on active sites. Such clusters may be electrically harmless if they remain separate, even in the presence of high humidity. Ceramics contain many active sites and are particularly prone to ionic conduction along damp external and internal surfaces.

Conduction by water sorbed on inorganic insulating surfaces is particularly conspicuous in the case of capacitors where the dielectric is a thin film of Al_2O_3 between aluminium electrodes. In this system the porosity of the dielectric as well as the moisture content of the atmosphere may be varied and fields of up to the order 10^6 V/cm may be applied. It has been shown[31] that water may be electrolysed in this system. The electrical properties at high field are particularly interesting.

The most complicated moisture problems arise with materials which are hydrogen bonded in the absence of water. Among these cellulose is the most important, since paper capacitors are widely used, often at high electrical fields. Paper consists largely of cellulose, a polymer related to sugars and starch which was classed as Group *D* in Chapter 11. Chapter 13, Section A gives the formula of a unit of a polymeric cellulose molecule. In solid cellulose or starch neighbouring polymeric chains are hydrogen bonded to each other, but these intermolecular bonds may be replaced by hydrogen bonding to water, which enters into the structure in amorphous regions of the polymer[29].

The more gross electrical effects of water sorption are exemplified in Fig. 6 which refers to starch where amorphous regions play a much greater part than in cellulose. Apart from the low frequency end of the curve which indicates direct conductivity, the figure illustrates two effects. At low moisture contents the high frequency dielectric relaxation of the moist starch occurs in the same frequency region as for the dry material. Here the adsorbed OH dipoles fluctuate in their directions together with the long-chain molecules to which they are attached (see also Fig. 7 in Chapter 13). With increasing

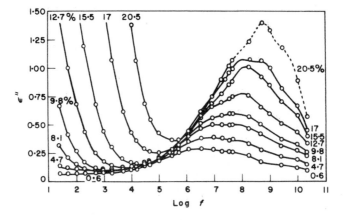

FIG. 6. Dielectric absorption in starch containing varying amounts of water (from Abadie et al.[32]).

moisture content the relaxation peak shifts gradually to higher frequencies, approaching gradually the relaxation frequency of water. For the highest moisture content the starch has swelled and liquid is present between the polymer chains.

Paper capacitors are made so that their moisture content is below the lowest value shown in Fig. 6, but even as low a value as 0.1% of sorbed water may be harmful[33]. The practical construction of a paper capacitor for use at high voltages is such that metal electrodes are interleaved with multiple layers of paper while the whole assembly is impregnated with a liquid dielectric—usually a hydrocarbon oil or a chlordiphenyl. Moisture adsorbed at the paper may be electrolysed at high fields and give rise to gas formation, while gas bubbles exposed to high fields lead to electrical discharges and thus break down. In the present context it is interesting that the amount of gas generated by the electrolysis of moist paper is greater than corresponds to the charge transport calculated on the basis of the observed average conductivity. The conductivity may possibly be enhanced within small regions where abnormally high stresses are present, an effect which seems plausible in view of the tendency of hydrogen bonding to form structures such as chains which respond cooperatively to an applied field.

REFERENCES

1. R. J. Meakins and R. A. Sack (1951). *Aust. J. scient. Res.*, **4**, 213.
2. B. V. Hamon and R. J. Meakins (1953). *Aust. J. Chem.*, **6**, 27.
3. R. J. Meakins (1962). *Trans. Faraday Soc.*, **58**, 478.
4. V. Daniel and J. W. H. Oldham (1961). *Trans. Faraday Soc.*, **57**, 694.
5. R. A. Sack (1952). *Aust. J. scient. Res.*, **5**, 135.
6. R. A. Sack (1963). *Trans. Faraday Soc.*, **59**, 1672.
7. V. Daniel and P. R. Vein (1964). *Trans. Faraday Soc.*, **60**, 1310, ERA report No. 5014.
8. V. Daniel (1958). *Trans. Faraday Soc.*, **54**, 1834.
9. V. Daniel (1964). *Trans. Faraday Soc.*, **60**, 1299.
10. J. G. Powles and J. A. E. Kail (1960). *Trans. Faraday Soc.*, **56**, 1996.
11. V. Daniel (1964). *J. theoret. Biol.*, **6**, 375.
12. P. J. Owston (1958). *Adv. Phys.*, **7**, 171.
13. L. Pauling (1960). "The Nature of the Chemical Bond." (3rd edition.) Cornell University Press, Cornell.
14. J. D. Bernal and R. H. Fowler (1934). *Proc. R. Soc.*, A, **114**, 1.
15. L. Onsager and M. Dupuis (1960). Rendicoti, S. I. Fermi X Corso, p. 294.
16. L. Onsager and M. Dupuis (1962). "Electrolytes", p. 27. Pergamon Press, London.
17. H. Gränicher, C. Jaccard, P. Scherrer and A. Steinemann (1957). *Discuss. Faraday Soc.*, **23**, 50.
18. R. P. Auty and R. H. Cole (1952). *J. Chem. Phys.*, **20**, 1309.
19. F. Humbel, F. Jona and P. Scherrer (1953). *Helv. phys. Acta*, **26**, 17.
20. H. O. Kneser, S. Magun and G. Ziegler (1955). *Naturwissenschaften*, **15**, 437.
21. H. Gränicher (1958). *Proc. Roy. Soc.*, A, **247**, 453.
22. H. Gränicher (1963). *Phys kond. Mat.*, **1**, 1.
23. M. Eigen and L. de Maeyer (1958). *Proc. R. Soc.*, **247**, 505.
24. M. Eigen, L. de Maeyer and H. C. Spatz (1964). *Ber. dt. Bunsenges. Phys. Chem.*, **68**, 19.
25. N. Bjerrum (1951). *K. danske Vidensk. Selsk. Skr.*, **27**, 1.
26. E. Spenke (1958). "Electronic Semiconductors" p. 321. McGraw-Hill, New York.
27. R. K. Chan, D. W. Davidson and E. Whalley (1965). *J. Chem. Phys.*, **43**, 2376.
28. J. A. Sussmann (1964). *Phys. kond. Mat.*, **2**, 146.
29. G. C. Pimentel and A. L. McClellan (1960). "The Hydrogen Bond." W. H. Freeman.
30. A. G. Day (1963). *Trans. Faraday Soc.*, **59**, 1218.
31. H. F. Church (1962). *Proc. Instn elect. Engrs*, **109B**, Suppl. 22, 399.
32. P. Abadie, R. Charbonniere, A. Gidel, P. Girard and A. Guilbot (1953). Colloque International du Centre National de la Recherche Scientifique, Paris, June 1953. CNRS, Paris.
33. Z. Krasucki, H. F. Church and C. G. Garton (1960). *J. electrochem Soc.*, **107**, 598.

CHAPTER 16

Ferroelectrics

A. INTRODUCTION

One might attempt to define a ferroelectric as a crystalline dielectric in which the Clausius–Mossotti catastrophe actually happens (see equation 2.59). That is, a dielectric where the electrostatic interaction at some temperature overcomes the tendency of thermal motion towards disorder. In such a dielectric dipoles would at low temperatures be ordered in some way which maximizes the energy of electrostatic interaction, while at higher temperatures this ordered state would give way to disorder among dipoles.

This definition is too wide in that it includes ordered structures other than ferroelectrics. In practice, it is so over-simplified as to be misleading. The energy of electrostatic interaction in ferroelectrics is only one of many energy terms which determine the cohesion of the solid. Many ferroelectrics do not contain identifiable permanent dipoles which become ordered. Ferroelectric crystals belong to several different structural types and the physics of ferroelectricity is a large subject, containing a great deal of intricate crystallography. This subject has been reviewed in the book by Jona and Shirane[1] in a comprehensive way. Later literature as well as earlier review literature is briefly summarized by Devonshire[2] from a thermodynamic viewpoint. The electrical properties of ferroelectrics have been reviewed by Merz[3] and, earlier, by Känzig[4]. Another earlier treatment by Megaw[5] is also helpful, particularly in explaining the crystallography. The present chapter aims to give a brief survey of the subject, with special reference to well documented examples of dielectric relaxation. The final section reviews, in more detail, a case not so well understood which has so far not been reviewed.

B. EXPERIMENTAL CRITERIA OF FERROELECTRICITY

All ferroelectric materials are crystalline. In the ferroelectric state they have a spontaneous polarization, which means that each unit cell of the crystal has a dipole moment. This is only possible if the crystallographic space group is polar, i.e. if the unit cell has no centre of symmetry. Otherwise each unit cell would contain for each dipole pointing in one direction another one, pointing in the opposite direction, and the resultant dipole moment would be zero.

Polar space groups are not restricted to ferroelectrics but characterize so-called pyroelectric crystals, whose polarization changes, in principle, with temperature. In order to be ferroelectric, a material must fulfil three conditions:

1. It must have a phase transition from a polar to a non-polar structure, or at least it must tend, with rising temperature, towards such a transition.
2. The polar phase must have a spontaneous polarization, that is the unit cell must actually have a dipole moment, not only belong to a space group which is capable of such a moment.
3. The direction of the spontaneous polarization must be reversible by an applied electric field. This third condition is the most important one.

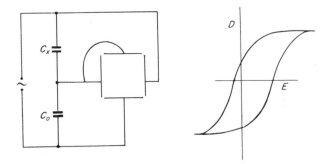

FIG. 1. Schematic Sawyer and Tower circuit with hysteresis loop as it would appear on the screen of the oscilloscope.

From the experimental point of view these criteria are not always easy to apply. Bogdanov[6] gives an authoritative analysis of the experimental tests which qualify a material as a ferroelectric. Some cases are ambiguous.

A typical ferroelectric exhibits hysteresis loops, in that D and E are related like B and H in a ferromagnetic material. Hysteresis loops may be observed on the screen of an oscilloscope. The Sawyer and Tower circuit which is generally used is shown schematically in Fig. 1, together with the hysteresis figure seen on the oscilloscope. Since the linear reference capacitor is in series with the specimen and $C_0 \gg C_x$, the voltage across C_0 is given by Q_x/C_0. This voltage, which is in phase with D in the specimen, is applied to the Y plates of the oscilloscope. The voltage applied to the X plates is in phase with the total externally applied voltage and thus with E in the specimen.

It should be noted that the hysteresis loop measured at mains frequencies represents properties at 50 c/s or 60 c/s and not the static properties of a ferroelectric. Normal, that is linear, dielectric relaxation contributes to the hysteresis loop. A linear dielectric gives on a Sawyer and Tower apparatus an elliptical trace the width of which measures the dielectric loss. At low frequencies this contribution to the width of the loop is often mainly due to the direct conductivity of the specimen. It is possible to compensate for direct conductivity by using modifications[5] of the basic circuit, so that a linear dielectric with ohmic conductivity but negligible a.c. loss gives a straight line on the screen of the oscilloscope.

The hysteresis loop shown by a ferroelectric has a shape as shown by the outline in Fig. 1. It is traced when an alternating voltage is applied. The response of the specimen to a constant direct voltage depends on its history, and the magnitude of the applied voltage.

The significance of a hysteresis loop is familiar in the context of magnetism. Hysteresis implies the existence of domains, that is regions within which dipoles (in the ferroelectric case polar unit cells) are parallel to one another. In equilibrium domains within a macroscopic specimen are arranged so that the resultant polarization of the specimen is zero in the absence of an external field. This condition represents the lowest energy state because a polarized specimen has an external field, in which energy is stored. This state of affairs is obvious in the magnetic case, where no free poles are present. In the electrical case, a macroscopic specimen with a resultant polarization need not have an external field if ions on the surfaces neutralize the polarization. However, this is an experimental complication, not a matter of principle.

The application of an external field has two consequences:

1. The polarization within each domain changes. At low field strengths this change is for a specimen in thermodynamic equilibrium linearly related to the applied field, and it is thus possible to define a dielectric constant, or, more precisely, a dielectric tensor, since polarizability depends on direction. For a polycrystalline specimen in equilibrium with random orientation of crystallites, measurement of ε at low field strengths yields an average value which is known as the low field permittivity. At not so low fields the polarization of a domain tends to vary non-linearly with field.

2. The configuration of domains changes if the applied field exceeds the coercive field for some domains and domains switch gradually or discontinuously into the direction of the field.

The features of the hysteresis loops may be understood as follows. The specimen which is depicted in Fig. 1 is being "cycled" at a field high enough to give saturation, indicated by the sharp corners of the loop. At the highest fields all domains have been switched into the field direction, so that an increase of the field is only able to increase the polarization within domains. When the field is reversed some domains retain their former alignment, so that the specimen has a remanent polarization at zero field. A field in reverse direction of sufficient magnitude (the coercive field) is needed to reverse these domains. The low field permittivity would be measured, ideally, for a virgin specimen of zero initial polarization.

It should be noted that dielectric hysteresis is not entirely analogous to magnetic hysteresis. The coercive field in magnetism is a fixed quantity but for a ferroelectric it depends on the magnitude and frequency of the switching field.

The area of a hysteresis loop represents an energy dissipation, and is thus a kind of dielectric loss. However, this loss will be mentioned only briefly, in Section H, since the switching of domains is a complicated subject in itself and has been reviewed recently[1,3,4]. The spontaneous polarization, i.e. the maximum polarization at high fields, varies greatly from ferroelectric to ferroelectric, between[1,2] the orders 10^{-7} coulombs/cm^2 and nearly 10^{-4} coulombs/cm^2.

C. FERROELECTRICITY AND ORDER–DISORDER TRANSITIONS

The sharp distinction made above between the dielectric constant of a single domain and the switching of domains is useful in practice but is not entirely justified. The orientation of domains can be affected by an applied field only because the ordered parallelism of displacement vectors within unit cells does not represent a very stable state of affairs. Some slight disorder exists within domains at all temperatures.

The concept of "order" is more obvious in ferromagnetism than ferroelectricity because well defined dipoles exist, in the magnetic spins of unpaired electrons. These dipoles interact with one another, and their mutual interaction causes a state of order, so that most dipoles within a domain are parallel. Not all are parallel because the influence of mutual interaction is opposed by the independent random motion of individual dipoles. When the random motion increases with rising temperature the mutual interaction is weakened. This constitutes positive feedback, and order declines in a catastrophic way with temperature until it collapses at a critical temperature or Curie point T_c.

The accelerated decline of order with increasing temperature can be derived in various ways. The most general and useful theory of order–disorder transitions is thermodynamic.

D. THE THERMODYNAMIC THEORY OF ORDER–DISORDER TRANSITIONS

The free energy of a gas is normally expressed as a function of two independent variables, which may be chosen from the following: volume, pressure, temperature, entropy, internal energy, or combinations of the above. An applied electrical or magnetic field, usually unimportant for a gas,

16. FERROELECTRICS

adds a third independent variable where such a field causes a not negligible change of the free energy.

In Chapter 3 the thermodynamics of a dielectric was treated as for a gas in which the electrical field is relevant, at constant volume. Fröhlich's[12] general treatment uses T and E as independent variables, while the bistable model was considered so that the independent variables were E and the degree of order ξ. The thermodynamics of a system with three independent variables is in general quite complicated. A survey with special reference to ferroelectrics is given by Devonshire[9].

The electrical polarization of a ferroelectric domain is in general connected with changes of shape and volume, that is mechanical strains and stresses in various directions. The most convenient choice of variables in this context is[1,9] to treat the electrical polarization P, the mechanical stress X and the temperature as independent variables. This leads in general to very complicated expressions because the polarization and the stress are vectors and the coefficients which describe their relationships to each other are tensors. This is explained by Jona and Shirane[1], in Section 5 of their introduction.

A simple expression in terms of the polarization is obtained if one formulates the free energy as a function of the stress and polarization and assumes that

1. all stresses are equal to zero, $X = 0$,
2. the polarization vector is directed along one of the crystallographic axes,
3. the non-polar phase has a centre of symmetry.

The appropriate free energy function for T, P and X as independent variables is the elastic Gibbs function defined by Devonshire[9] as G_1. The condition $X = 0$ signifies that G_1 is a function of T and P only. The symbol for this free energy is chosen variously by different authors. In any case, the function is used as an expansion in terms of P^2 which here shall be written

$$G(P, T) = G(P, T, X)_{X=0} = G_0(T) + \tfrac{1}{2}A(T)P^2 + \tfrac{1}{4}B(T)P^4 + \cdots. \quad (16.1)$$

When this expression has only positive coefficients multiplied by powers of P it has always a minimum for $P = 0$, as in normal, paraelectric dielectrics. However, if $A(T)$ may change sign, the polarization is capable of equilibrium values $P_e \neq 0$, and ferroelectric properties are possible.

The dielectric response may in general be deduced from the relation[1,9]

$$\left(\frac{\partial G}{\partial P}\right)_{TX} = E. \quad (16.2)$$

An analogous relation is deduced here for constant temperature and volume (equations 16.6–16.12).

Differentiation of equation (16.1) gives

$$AP + BP^3 + \cdots = \overset{\circ}{E}. \qquad (16.3)$$

This equation yields the incremental value of the susceptibility

$$\chi = \left(\frac{\partial P}{\partial E}\right)_{TX} \qquad (16.4)$$

which gives an incremental value of the dielectric constant by way of the general formula

$$\frac{\varepsilon - 1}{4\pi} = \chi. \qquad (16.5)$$

Equations (16.4) and (16.5) refer to a specimen in thermodynamic equilibrium at constant temperature, and signify that an applied small field changes the equilibrium polarization P_e by a small amount. These equations are unambiguous only if the field due to the specimen in the space surrounding it is known. Since this field influences the free energy of the solid, it influences the coefficients in equation (16.1) and the value of ε depends thus on the shape and surroundings of a given specimen. This consideration is neglected in the thermodynamic treatment, where the free energy refers to a specimen of infinite size. It will be seen in the next section that the error introduced by this neglect is not as important in practice as might appear at first sight.

The thermodynamic treatment so far provides no reason why $A(T)$ should have negative or positive values. Such a reason can only follow from a structural model. It will be shown that the bistable model with electrostatic interaction leads to a temperature dependence of $A(T)$ such that this coefficient goes through zero at a critical temperature, and that the model approximates ferroelectric behaviour.

In Chapters 3 and 4 the free energy $F(E, \xi)$ of the bistable model has been derived in terms of the order variable ξ which is simply related to P (it is proportional to P according to equation 3.22 if $\varepsilon_\infty = 1$). Devonshire[9] also derives $F(E, P)$ for a somewhat more sophisticated form of the model which includes electronic polarization. These treatments can be used to derive ε and the coefficient $A(T)$ in equation (16.1), but the procedure is somewhat involved. While the function $G(P, T)$ is defined for constant stress, $F(P, E)$ is defined for constant volume. We shall here use a derivation given by Burgess[10,11], since this is particularly convenient from the viewpoint of fluctuations.

The polarization P is defined per unit volume but for the treatment of fluctuations it is important not to obscure the role of the volume so that the

16. FERROELECTRICS

following derivation is in terms of $M = Pv$. In general, the Helmholtz free energy is

$$F = U - TS \tag{16.6}$$

and its total differential

$$dF = dU - T\,dS - S\,dT. \tag{16.7}$$

Now (see equations 3.3 and 3.4) a dielectric in thermal equilibrium with a large heat reservoir of temperature T changes its heat content by

$$dU = T\,dS + \frac{E}{4\pi}\,dD \tag{16.8}$$

when a field is applied. For the process in question $S\,dT$ is zero because the temperature is kept constant. The dielectric displacement D is in general given by equation (2.8), but for a ferroelectric $\varepsilon_s \gg 1$ so that one may put in good approximation

$$D = 4\pi P. \tag{16.9}$$

For a dielectric of constant volume v this gives

$$dU = T\,dS + vE\,dP$$
$$= T\,dS + E\,dM \tag{16.10}$$

and equation (16.7) transforms to

$$dF = E\,dM \tag{16.11}$$

for the conditions defined. Now (see also equation 16.2)

$$\left(\frac{\partial F(M, E)}{\partial M}\right)_{T,v} = E \tag{16.12}$$

while the incremental susceptibility follows from

$$\frac{1}{\chi} = \left(\frac{\partial E}{\partial P}\right)_{T,v} \tag{16.13}$$

In view of equation (16.12) the incremental susceptibility is thus given by

$$\frac{1}{\chi} = v\left(\frac{\partial^2 F}{\partial M^2}\right)_{T,v,M=M_e} \tag{16.14a}$$

$$= \left(\frac{\partial^2 F}{\partial P^2}\right)_{T,v,P=P_e} \tag{16.14b}$$

where the last subscript signifies that the differential is taken at the equilibrium value of M or P.

The free energy for the bistable model with electrostatic interaction has been deduced in Chapter 3, in terms of ξ. It was assumed that there was no spontaneous polarization, i.e. $\xi_e = 0$ for $E = 0$, and furthermore $\xi \ll 1$. Differentiation of equation (3.26) for the free energy gives

$$\left(\frac{\partial^2 F}{\partial \xi^2}\right)_{T,v} = 4Nk(T - T_c) \tag{16.15}$$

where T_c is given by equation (3.28). This expression can be expressed in terms of the polarization P by using equation (3.22) when

$$\left(\frac{\partial^2 F}{\partial P^2}\right)_{T,v} = \frac{1}{4\mu^2 N^2} \cdot \left(\frac{\partial^2 F}{\partial \xi^2}\right)_{T,v}$$

$$= \frac{4\pi}{3} \cdot \frac{T - T_c}{T_c}. \tag{16.16}$$

This gives on the basis of equation (16.14b)

$$\frac{1}{\chi} = \frac{4\pi}{3} \frac{(T - T_c)}{T_c} \quad \text{for} \quad T > T_c. \tag{16.17}$$

The susceptibility calculated from equation (16.17) may be compared with that calculated from equation (16.3) for the same conditions, i.e. for the case where higher orders in P may be neglected. Differentiation according to (16.4) gives

$$A(T) = \frac{1}{\chi(T)} \quad \text{for} \quad T > T_c \tag{16.18}$$

$$= \frac{4\pi}{3} \cdot \frac{(T - T_c)}{T_c}. \tag{16.19}$$

The value of χ below the critical temperature has to be calculated from the increment of the polarization above $P_e \neq 0$ which is induced by a small field E. It is found to be[1,9,10]

$$\chi = 2 \frac{4\pi}{3} \frac{(T - T_c)}{T_c} \quad \text{for} \quad T < T_c. \tag{16.20}$$

The function $A(T)$ obviously retains its form given by (16.19) throughout the temperature range.

Figure 2, adapted from Devonshire[9] shows the dependence of the free energy on P, at three constant temperatures of which the middle one is T_c. It also shows the temperature dependence of P. The function $P(T)$ is similar to that shown for a ferromagnetic material. Indeed, the bistable model treated is one of the simplest models for an order–disorder transition. This may also be appreciated by considering equation (3.26). The entropy term which is proportional to ξ^2 and represents an energy of random motion is at T_c neutralized by an energy term due to electrostatic interaction, so that at that temperature disorder changes to order.

The bistable model, with fixed dipoles of moment μ which can be ordered, is not in general a good model of ferroelectricity, as will be seen in Section G.

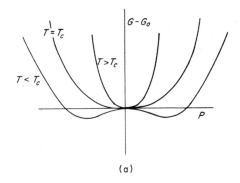

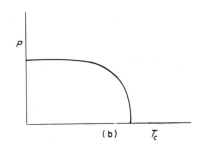

FIG. 2. Schematic representation for the bistable model of (a) the free energy as a function of the polarization and (b) the polarization as a function of temperature.

However, the free energy expression calculated from the model is a useful starting point.

The free energy equation (16.1), with $A(T)$ given by an expression of type (16.19), is a very good approximation for the group of ferroelectrics typified by triglycine sulphate—$(NH_2CH_2COOH)_3H_2SO_4$, abbreviated T.G.S. The properties of T.G.S. can be represented by an equation

$$G(P, T) = G_0(T) + \frac{2\pi}{C}(T - T_c)P^2 + \tfrac{1}{4}BP^4 + \tfrac{1}{6}BP^6. \quad (16.21)$$

The constants contained in the terms with P can be derived from various electrical measurements[1]. If the term independent of P varies little with temperature near T, then the electrical data must also predict correctly the dependence of the specific heat on temperature. The measured and calculated values for c_p near the transition temperature of T.G.S. agree quite well[1]. This is rather remarkable when one bears in mind that the heat of transition, calculated from the area under the $c_p(T)$ curve is only 150 cal/mole. The cohesive energy of a solid is large compared with this value, and factors other than the polarization might easily change the temperature dependence of the free energy so as to spoil agreement with the electrical data.

The phase transition at the critical temperature of T.G.S. is of the second order, in that both U and S change continuously through the transition. Many transitions in ferroelectrics are of the first order[1], with discontinuous changes of U and S. For a second order transition the dielectric constant is (theoretically) infinite at T_c. For a first order transition there are two values of $\varepsilon(T_c)$, approached from above and below the transition respectively.

E. FLUCTUATIONS NEAR THE CRITICAL TEMPERATURE AND THE ROLE OF FIELD THEORY

The physical significance of the high susceptibility below as well as above T_c is best considered in terms of fluctuations. In Chapter 4 it has been shown that for the bistable model, in the absence of electrostatic interaction, χ and the fluctuation $\overline{M^2}$ are simply related. In that chapter this argument was not pursued to include electrostatic interaction since the interrelation between microscopic, molecular properties and the fluctuation of the macroscopic polarization is very complicated, as explained by Fröhlich[12]. However, the thermodynamic treatment of ferroelectrics is a macroscopic treatment and insofar as the free energy of a macroscopic specimen can be expanded as a function of P^2 the relationship between χ and the fluctuation of the macroscopic polarization is unambiguous.

According to equations (4.10) and (4.11), we have in general (writing M_e for $\bar{x}$) for a specimen of volume v

$$\overline{(M - M_e)^2} = \frac{kT}{\left(\dfrac{\partial^2 F}{\partial M^2}\right)_{M=M_e}}. \tag{16.22}$$

Comparison of this equation with equation (16.14a) gives

$$\overline{(M - M_e)^2} = kT\chi v. \tag{16.23}$$

The fluctuation of M is in principle measurable. However, the conditions of measurement become important where electrostatic interaction is appreciable, as they are important also for the measurement of the dielectric constant. χ, as well as the fluctuations of M, now depend on the shape of the specimen and its surroundings, while equation (16.23) is generally valid.

16. FERROELECTRICS

The role of shape and surroundings becomes clear when one compares Fröhlich's equations (8.22 and 8.23) for the fluctuation of the moment of a sphere in vacuo and of a sphere embedded in its own medium. When ε_s tends towards infinity the fluctuation for the sphere in its own medium also tends towards infinity. The fluctuation for the sphere in vacuo stays small, and tends towards a constant for $\varepsilon_s \to \infty$. This does not contradict either equation (16.23) or the definition of ε_s. It signifies a change of the free energy of the solid due to the external field. For the sphere in vacuo the fluctuations of the

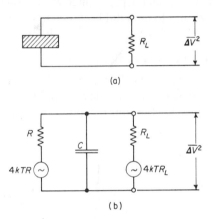

FIG. 3. Circuit for the measurement of fluctuations of the polarization of (a) real circuit and (b) an equivalent circuit (from Brophy[13]).

polarization cause a fluctuating field in space, and if the fluctuation were to increase as $\varepsilon_s \to \infty$ the field energy in space would become infinite too. Hence the external field of the fluctuation acts as a negative feed back, and suppresses the fluctuations.

When ε_s is measured in the normal way, the dielectric specimen carries metallic electrodes. If these electrodes envelop the specimen they eliminate the external field of the fluctuations. If two electrodes are interconnected by a measuring instrument the external field is reduced, and in the case of measurements for ferroelectrics by bridge or Sawyer and Tower methods, the measured dielectric constant corresponds fairly well to the free energy equation for a specimen of infinite size. Errors are introduced chiefly by poor contacts, and the not negligible magnitude of the applied field.

When the fluctuation of the polarization of a ferroelectric specimen is measured, the specimen cannot supply power, only a potential which may be amplified by a very sensitive voltmeter. The circuit used in practice is shown in Fig. 3(a). The voltage across the terminals of the ohmic resistance R_L is analysed by Brophy in terms of the equivalent circuit (Fig. 3b). The resistance

R signifies a value characteristic for a frequency interval $\omega \pm \Delta\omega$ which is determined by a filter in the voltmeter.

Brophy uses

$$a\omega^2 \langle \Delta P^2 \rangle = \frac{4kT_s}{R} \qquad (16.24)$$

where T_s is the specimen temperature and $\langle \Delta P^2 \rangle$ is the spectral density of the polarization fluctuation for the frequency interval in question. The voltage fluctuation measured by the meter depends on R, R_L, C, ω and the temperature T of R_L. The value of R_L has to be suitably chosen to give maximum sensitivity. Values of the order 10^{-15} V²/cycle are recorded[13].

F. RELAXATION IN TRIGLYCENE SULPHATE

The frequency dependence of the fluctuation of P can be analysed in terms of the power spectrum, which is for a single relaxation time given by equation (4.38). Brophy[13] measures fluctuations in a range of temperatures very close

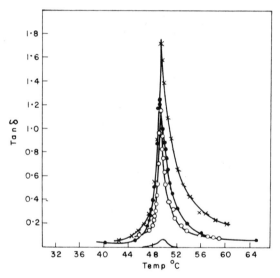

FIG. 4. tan δ versus temperature in T.G.S. for different frequencies (from Lurio and Stern[14]).

to T_c and finds that they are for a given temperature consistent with a single relaxation time. He finds $\tau = 5\cdot3 \times 10^{-3}$ sec below T_c and $2\cdot7 \times 10^{-3}$ sec above T_c.

The measurement of relaxation times by way of fluctuations is equivalent to their measurement by more usual methods, but extrapolated to zero applied field. Relaxation in T.G.S. has been measured using conventional methods. Figure 4 shows tan δ as a function of frequency and temperature[3,14].

The peak values are at T_c and, from the data shown, the relaxation time at T_c would appear to be much shorter than that deduced by Brophy. This discrepancy is mitigated somewhat if one bears in mind the field dependence[15] of tan δ shown in Fig. 5, but remains unexplained. The distribution of relaxation times in T.G.S. is temperature dependent. Hill and Ichiki[15] find that $y(\tau)$ can be described by a Gaussian distribution which narrows as T_c is approached. The median relaxation time becomes longer, as $\tau = \text{const.}/T - T_c$.

At this stage we may enquire to what extent the relaxation behaviour follows from the equation for the free energy. In terms of irreversible thermo-

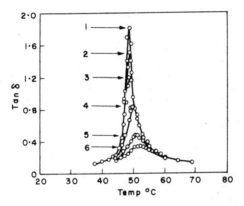

FIG. 5. tan δ versus temperature in T.G.S. at 2600 Mc/s for different d.c. bias fields (from Hill and Ichiki[12]). Curve 1–Bias 0; similarly, 2–0·6; 3–1·2; 4–2·4; 5–6·4; 6–1·06 kV/cm.

dynamics the median relaxation time can be calculated from $F(P)$. For the bistable model, with electrostatic interaction, equations (3.26) and (3.31)–(3.33) give

$$\tau = \frac{K}{4k(T - T_c)} \quad (16.25)$$

which is the relationship found experimentally for T.G.S., a material whose thermodynamic properties resemble the bistable model.

The narrowing of the distribution $y(\tau)$ on approach to T_c does not follow from the thermodynamics without further assumptions. However, there seems to be a close connection: Hill and Ichiki[15] report that the half width of the Gaussian distribution of relaxation times changes as $1/T - T_c$.

While T.G.S. is the simplest ferroelectric from the thermodynamic point of view, in that it resembles the bistable model, its structure is quite complicated[1]. The structural information to be gained from T.G.S. is, in the present context, that extensive disorder is possible in a solid structure where the unit cell is polar below T_c. This means that a fair number of unit cells in the solid must point in the "wrong" way, and this does not cause prohibitive stresses. Rotating dipoles could hardly behave in this way. The most probable[1,2] mechanism for dipole reversal in T.G.S. is a movement of a hydrogen atom between two positions, accompanied by movements of the glycine ions.

G. THE PEROVSKITE GROUP AND LATTICE DYNAMICS

The structural significance of dipole reversal is more easily seen in the perovskite group of ferroelectrics which is the most important group from the practical point of view though thermodynamically less simple. The best known member of this group is $BaTiO_3$ which is cubic above $T_c = 120°C$ and tetragonal below this transition temperature. Other transitions to structures of still lower symmetry occur at lower temperatures. The closely related $SrTiO_3$ has the cubic perovskite structure down to low temperatures[1].

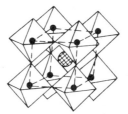

FIG. 6. Ideal perovskite structure ABO_3 considered as a framework of BO_6 octahedra (from H. Megaw; 1946. *Trans. Faraday Soc.*, **42A**, 224). ○ A^{++} ion; ● B^{++++} ions, oxygens at corners of octahedra.

Figure 6 shows the ideal cubic perovskite structure. For cubic $BaTiO_3$, titanium ions (small solid circles) are in the centre of oxygen octahedra. The feature responsible for the ferroelectric transition of $BaTiO_3$ is shown in Fig. 7, which represents the oxygen environment of a titanium ion in the tetragonal form, which is a slight distortion of the cubic form. While titanium is centrally situated in the cubic form it is off centre in the tetragonal form. The displacement of the centre of gravity of oxygens against titanium is the main reason for the reversible dipole moment of a unit cell of $BaTiO_3$. The figure shows why the moment is reversible, and illustrates that the smaller the displacement the more easily will the dipole reverse.

One might attempt to describe the dipole reversal in terms of a potential well model, by considering that there are two positions of minimum potential energy available for the titanium ion in Fig. 7, one near the lower and one near the upper oxygen. These wells could approach each other as the temperature of the tetragonal phase is raised, and merge at T_c. The state of affairs above T_c would then be described by a single potential well, which near T_c is very wide. This would signify a very small force constant of thermal vibrations just above T_c. However, while this model is easy to visualize it is so crude as to be misleading. It neglects the fact that in a crystal the framework surrounding the ion moves as well as the ion itself.

Cochran[16] gives a quantitative treatment of the thermal vibrations in a ferroelectric crystal which is based on lattice dynamics (see Chapter 9). This theory explains ferroelectricity on the basis of a vanishing force constant, as in the crude model sketched below. However, the force constant now refers

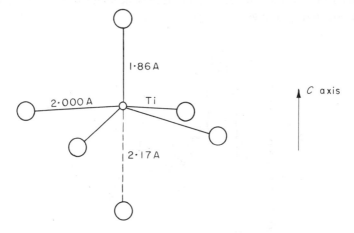

FIG. 7. The oxygen ions surrounding a titanium ion in the tetragonal form of BaTiO$_3$ (from Jona and Shirane[1]).

to a transverse mode of vibration of the crystal lattice as a whole. The simplest kind of mode which is considered may be visualized most easily for a simple structure like that of NaCl. We may imagine that all Na$^+$ ions, in a rigid array, move against the rigid array formed by the Cl$^-$ ions. This movement may occur for instance in the [001] direction while the elastic wave propagates in the [100] direction.

Cochran reviews the theory of the dielectric constant of ionic crystals. As pointed out by Szigeti[17], the usual approach of drawing a sphere around a dipole and separating the field into an internal and external component is not applicable. Ions overlap and their oscillations are too strongly coupled for an individual ion to be considered as an individual oscillator. However, it is possible to consider as independent oscillators the normal vibrations of the crystal. Vibrations at infra-red frequencies imply both atomic and electronic displacements. Electronic and atomic contributions add for transverse infrared modes and subtract for longitudinal ones. For an individual transverse vibration of frequency ω the contribution to the dielectric constant[16,17] is

$$\varepsilon_s - \varepsilon_\infty = \frac{4\pi(\varepsilon_\infty + 2)^2(Z'e)^2}{9v\mu\omega^2} \quad (16.26)$$

where ε_∞ is here used to denote an optical dielectric constant, $Z'e$ is an effective electronic charge, v the unit cell volume and

$$\mu = \frac{m_1 m_2}{m_1 + m_2} \quad (16.27)$$

m_1 and m_2 being the masses of the two kinds of ions which vibrate in relation to each other in the transverse mode. It may be seen that the contribution of the mode in question goes to infinity if ω goes to zero.

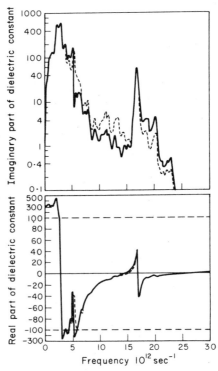

FIG. 8. The real and imaginary part of the dielectric constant of SrTiO$_3$ at 300°K calculated for two variants of Cowley's model (from Cowley[18]).

Cochran shows that the frequency of a transverse vibration may tend to zero if the vibration is not harmonic, that is if restoring force and displacement are not strictly proportional. In first approximation his result may be summarized by stating that for one particular transverse mode

$$\omega^2 = \text{const.}(T - T_c). \quad (16.28)$$

This expression, substituted in (16.26), implies a second-order ferroelectric transition.

The refinements of the dynamic theory are too complex to be discussed here in detail. Cowley[18] reviews earlier work and formulates a model such that the temperature dependence of the normal modes of vibration arises from the anharmonic interactions between those normal modes. This formulation allows the derivation of macroscopic dielectric properties, in particular of the free energy, from the microscopic data of individual ferroelectrics. By using a model of $SrTiO_3$ which includes the anharmonic interactions between the normal modes in a simple way, Cowley deduces the frequency dependence of the dielectric constant and loss. Figure 8 shows the result of the calculations for two variants of the model. Cowley evaluates the reflectivity which follows from the model and finds satisfactory agreement with experiment for $SrTiO_3$ at room temperature.

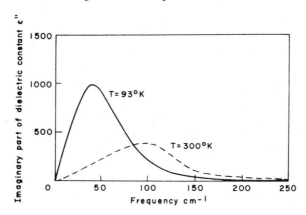

FIG. 9. The imaginary part of the dielectric constant for $SrTiO_3$ deduced from measurements[16] of the reflectivity of 300°K and 93°K (from Barker and Tinkham[19]).

Cowley's treatment does not necessarily lead to a single "ferroelectric" mode. Under certain circumstances several normal modes may participate in the transition. However, the temperature dependence of relaxation may still be described by Cochran's model. Figure 9 shows[19] the dependence of $\varepsilon''(\omega)$ on the temperature for $SrTiO_3$. It may be seen that a broad maximum of $\varepsilon''(\omega)$ shifts towards lower frequencies on approaching T_c and becomes somewhat sharper. Similar results[20] have been obtained for $BaTiO_3$ at temperatures above T_c. The experimental data available are not yet as detailed as those calculated from Cowley's model.

Cowley[18] discusses the approximations involved in the theories of ferroelectricity. The thermodynamic theory is of course only phenomenological in that it is not concerned with the structure of the material it describes, while the dynamic theory can be fitted to the structural data and gives structural insight.

H. RELAXATION BY MOVEMENTS OF DOMAIN BOUNDARIES

So far we have considered dielectric loss within a single domain. In practice, the movements of domain walls may be important, and the energy loss due to movements of domains can be very large compared with that due to loss within a domain.

The mechanisms by which domains increase or decrease their size have been reviewed by Merz[1,3] and are very complicated. The timing of these mechanisms is more complicated still and cannot be considered here. However, it seems worth while to discuss whether there are cases where the volume and shape of domains fluctuates spontaneously, so that the distinction between low field and hysteresis losses disappears.

In the case of ferromagnetism the extent to which domains are fixed is measured by the coercive field, and the magnitude of this field is related to the energy stored in a wall between domains. Such a wall may be considered to be many unit cells thick, and the directions of spins may vary continuously throughout the wall. The ferroelectric case is more difficult[1,3], since mechanical stresses are more important. Dipoles pointing at odd angles are hardly credible, but a change of the polarization of a unit cell with distance may be an acceptable assumption. Calculations of the energy of domain walls in $BaTiO_3$ at room temperature result in values of the order 1 to 10 ergs/cm^2. This is not a high value, since it is at most equal to an energy of the order kT/unit cell in the wall. The coercive field has been estimated from the data on wall energy but experimentally found coercivities are systematically too low compared with theory.

In practice, domain movements in ferroelectrics far from their transition temperature do not normally contribute significantly to the low field dielectric constant or loss. This means that the shape and size of domains does not fluctuate spontaneously, or fluctuates much more slowly than the polarization within domains. However, as the direct measurement[13] of fluctuations in T.G.S. shows, the distinction between fluctuations within domains and fluctuations of domains tends to vanish near the critical point. Besides[13,15] the distribution of relaxation times narrows towards T_c, suggesting a single process at the critical temperature of a second order transition.

I. RELAXATION FERROELECTRICS

While the distinction between hysteresis losses and low field losses in ferroelectrics is normally blurred only quite close to a transition temperature, there exists a group of materials where such conditions persist over a wide temperature range. This group is made up of mixed oxides of perovskite or related structure and will be denoted as "relaxation ferroelectrics".

Materials belonging to the group in question were first investigated by Skanavi and co-workers[21], with special reference to their nearly linear electrical response combined with a high permittivity. Smolenski[22] and co-workers investigated a wider range of materials, under the heading of ferroelectrics with unsharp phase transitions. It is difficult to define a group of relaxation ferroelectrics in an exact way, since unsharp phase transitions can be due to macroscopic inhomogeneities. However, it seems now clear that at least some of the materials in question are homogeneous in every sense in which this term is normally used[23].

The best known relaxation ferroelectrics are[21] solid solutions of the composition $(1-x)\mathrm{SrTiO_3} + \frac{x}{3}\mathrm{Bi_2O_3}.3\mathrm{TiO_2}$ (for short B.S.T.). These solid solutions have the cubic structure of $\mathrm{SrTiO_3}$, but three Sr^{++} ions are replaced by two Bi^{+++} ions and a vacancy, all in positions normally occupied by strontium. This structure is stable[24] up to at least $x = 0.3$, over a wide temperature range. The Smolenski group claims[23] B.S.T. as a ferroelectric with unsharp phase transition.

Figure 9 shows the dielectric data of a single crystal of B.S.T. with $x = 0.144$ as a function of temperature and frequency[25]. While ε' rises toward low temperatures in a way reminiscent of a ferroelectric, the peak value of tan δ depends on frequency as in a normal dielectric. The average relaxation time is very long compared with $\mathrm{SrTiO_3}$ and the distribution of relaxation times is very wide. The field dependence of the permittivity in B.S.T. is slight compared to a normal ferroelectric of comparable permittivity. The Sawyer–Tower figures[25,26] have the appearance characteristic for a normal dielectric up to fields of several times 10^4 V/cm. Single crystals show a very slight birefringence and electro-optical effects which suggest[25] a slight polarization within regions of sub-microscopic size.

Gubkin et al.[26] examine a wide range of compositions of B.S.T. and find that at constant frequency the peak values of tan δ and ε as a function of temperature shift systematically to lower temperatures as x decreases. The dependence is not linear in x. Bogdanov et al.[27] investigate the system $(1-x)\mathrm{BaTiO_3} + \frac{x}{3}\mathrm{Bi_2O_3}3.\mathrm{TiO_2}$ and examine the dependence of properties on x. They find that dielectric relaxation effects are observed above the ferroelectric Curie points of the mixtures for value of x beyond a certain minimum. This behaviour agrees quite well with the observations[26] on B.S.T. In the barium material a kind of unsharp phase transition is apparent at a bismuth concentration of 0.16%, a concentration near the value characteristic for the onset of relaxation phenomena in B.S.T. as well. Bogdanov[27] suggests an electronic mechanism for the relaxation behaviour of the barium bismuth titanates.

The only quantitative theory of the relaxation in B.S.T. or any other relaxation ferroelectric has been proposed by Skanavi[28]. This theory is difficult to follow but has been briefly summarized by Cross[29]. Skanavi assumes that the bismuth ions and vacancies distort neighbouring oxygen octahedra, producing more than one "off centre" equilibrium position for some titanium ions. The disturbed titanium ions hop between these positions

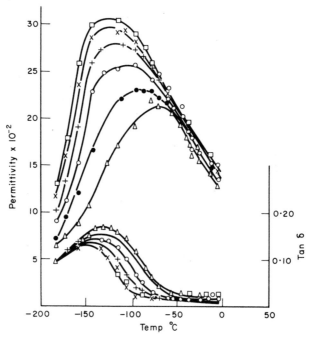

FIG. 10. Permittivity and loss angle as a function of temperature and frequency on a single crystal of B.S.T. with $x = 0.144$ (from Cross[29]). △—500 Kc/s; ●—100 Kc/s; ○—10 Kc/s; +—1 Kc/s; ×—100 c/s; □—30 c/s.

and somehow constitute "quasi dipoles". Skanavi assumes that these dipoles exist in a medium of high and rapidly relaxing polarizability, and are subject to a field which, in the notation of the present monograph, is given by

$$E_i = E + \beta_\infty P_\infty + \beta_D P_D \qquad (16.29)$$

which enters the relaxation equation for P_D, as in equation (2.63). The factors β_∞ and β_D are treated as parameters which may be fitted to experimental data. It is found that the observed data demand a negative value for β_D. That is, they imply a dipolar interaction such that it tends to reduce the parallelism caused by $\beta_\infty P_\infty$. The negative β_D is demanded[29] by data on $\varepsilon'(T)$ and $\varepsilon'(E)$ as well as the magnitude of the average relaxation time.

Equation (16.29) is not exact. It is a relatively trivial point that this assumption for the inner field neglects the circumstance that the positive feedback via ε_∞ is non-linear (see Chapter 2, Section C). However, a constant value for β_D is not obvious and would have to be deduced from a rigorous treatment[12] of electrostatic interaction. Furthermore, it is difficult to assess how the presence of Bi^{+++} and vacant sites affects the thermal vibrations of the basic $SrTiO_3$ lattice. The role of β_∞ and P_∞ is difficult to disentangle from that of β_D and P_D. Skanavi's treatment is interesting as, in principle, a simple mathematical statement of a very unusual situation.

Smolenski and co-workers, in a series of papers reviewed by Isupov[23], consider relaxation ferroelectrics which are single phase thermodynamically as being nevertheless inhomogeneous on an extremely small scale. Isupov speaks of "domainoids" which have widely varying values of T_c so that over a wide range of temperature some are above, some below a ferroelectric transition. The boundaries of these domainoids are assumed to fluctuate spontaneously. The size of domainoids may be of the order 100 Å. They are assumed to differ from each other in composition, i.e. their differences are based on statistical fluctuations in the distribution of ions.

The ultimate structural reasons for the dielectric behaviour of relaxation ferroelectrics are still not fully understood. Some of the materials in question are clearly ferroelectric, while others do not differ much from normal dielectrics. The common characteristic of the group is a blurring of the distinction between fluctuations of the polarization of units of atomic dimensions, and of units which are larger. This blurring could be due to different reasons for different materials.

REFERENCES

1. F. Jona and G. Shirane (1962). "Ferroelectric Crystals." Pergamon Press, London.
2. A. F. Devonshire (1964). *Rep. Prog. Phys.*, **27**, 1.
3. W. J. Merz (1962). *Prog. Dielect.*, **4**, 101.
4. W. Känzig (1957). *Soviet Phys. solid St.*, **4**, 1.
5. H. D. Megaw (1957). "Ferroelectricity in Crystals." Methuen, London.
6. S. V. Bogdanov (1963). *Soviet Phys. solid St.*, **5**, 591 (originally in 1963). *Fizika tverd. Tela*, **5**, 811.
7. C. B. Sawyer and C. H. Tower (1930). *Phys. Rev.*, **35**, 269.
8. H. Diamant, K. Drenk and R. Pepinski (1957). *Rev. scient. Instrum.*, **28**, 30.
9. A. F. Devonshire (1954). *Adv Phys.*, **3**, 85.
10. R. E. Burgess (1958). *Can. J. Phys.*, **36**, 1569.
11. R. E. Burgess. Private communication.

12. H. Fröhlich (1958). "Theory of Dielectrics." Oxford University Press, London.
13. J. J. Brophy (1965). "Fluctuation Phenomena in Solids" (R. E. Burgess ed.), p. 1. Academic Press, London and New York.
14. A. Lurio and E. Stern (1960). *J. appl. Phys.*, **31**, 1125.
15. R. M. Hill and S. K. Ichiki (1963). *Phys. Rev.*, **132**, 1603.
16. W. Cochran (1960). *Adv. Phys.*, **9**, 387.
17. B. Szigeti (1949). *Trans. Faraday Soc.*, **45**, 155.
18. R. A. Cowley (1965). *Phil. Mag.*, **11**, 673.
19. A. S. Barker and M. Tinkham (1962). *Phys. Rev.*, **125**, 1527.
20. J. M. Ballantyne (1964). *Phys Rev.*, **136**A, 429.
21. G. I. Skanavi and N. Matveeva (1956). *Zh. Eksp. teor. Fiz.*, **30**, 1047.
22. G. A. Smolenski, V. A. Isupov, A. I. Agranovskaya and S. N. Popov (1960). *Soviet Phys. solid St.*, **11**, 2906.
23. V. A. Isupov (1964). *Izv. Akad. Nauk. tadzhik. SSR.*, **20**, 560.
24. L. E. Cross (1962). *Proc. Instn elect. Engrs*, **109**B, Suppl. 22.
25. L. E. Cross (1962). E.R.A. report L/T415.
26. A. N. Gubkin, A. M. Kashtanova and G. I. Skanavi (1961). *Fizika. tverd. Tela*, **3**, 1110.
27. S. V. Bogdanov, K. V. Kiseleva and V. A. Rassushin (1965). *Soviet Phys. Crystallogr.*, **10**, 58 (transl.).
28. G. I. Skanavi (1958). "Halbleiter und Phosphore." (M. Schön and H. Walker eds.). Vieweg, Braunschweig.
29. L. E. Cross (1960). E.R.A. report L/T399.

APPENDIX I

Units and Constants

Three systems of units may be used for the purposes of dielectric relaxation. Namely, the absolute electrostatic system (denoted by e.s.u.) which is based on three basic units (centimetre, gram and second for length, weight and time) and two practical systems, which have for electrical purposes four basic units, namely length, weight, time and an electrical unit. The difference between the two practical systems is that the practical c.g.s. system uses the centimetre and gram for length and weight, while the practical m.k.s. system (or international system of S.I. units) uses the meter = 100 cm and the kilogram = 1000 grams. All three systems use the second as time unit.

The dielectric constant is dimensionless in e.s.u. but has a physical dimension in the practical systems. One then writes in the practical systems

$$\varepsilon(\text{practical}) = \varepsilon_0 \varepsilon \tag{I.1}$$

where ε represents the dimensionless "relative" dielectric constant which is appropriate to the electrostatic system. The constant ε_0 signifies the dielectric constant of empty space and has the dimension Farad/length in either practical system but different numerical values in c.g.s. and m.k.s. units. Similar conversion factors apply to other entities relevant for dielectric relaxation, as set out below. In several cases more than one symbol for a given entity is in current use. The symbol used in this monograph is given first.

It may be noted that the factor 4π which enters into the conversion factor for ε_0 is due to the different definition of D in the electrostatic system and the two practical systems.

Table A gives some constants which are relevant for dielectric relaxation.

TABLE A

Name	Symbol	Magnitude and Dimension
Charge of the electron	e	$4\cdot803 \times 10^{-10}$ e.s.u.
Boltzmann constant	k	$1\cdot380 \times 10^{-16}$ erg/°C
Planck's constant	h	$6\cdot625 \times 10^{-27}$ erg sec
Avogadro's number	N	$6\cdot025 \times 10^{23}$
Faraday constant	eN	96522 coulombs
Energy units:		
1 kcal = 10^3 cal		$4\cdot19 \times 10^{10}$ ergs
1 eV/mol		$9\cdot65 \times 10^{11}$ ergs
Unit of dipole moment	1 Debye	$4\cdot803 \times 10^{-18}$ e.s.u.

1 eV/mol = 23,030 kcal/mol; 1 mol contains N atoms or molecules, where N is Avogadro's number.

TABLE B

Entity	Symbols used in literature	Dimension in e.s.u.	Units in practical c.g.s. and m.k.s.	Multiply data in e.s.u. by conversion factor to get data in practical c.g.s.	Multiply data in e.s.u. by conversion factor to get data in practical m.k.s.
Dielectric constant of free space	ε_0, κ_0	dimensionless	$\dfrac{\text{Farad}}{\text{length}}$	$\dfrac{1}{4\pi \times 9 \times 10^{11}} = 0.885 \times 10^{-13}$	0.885×10^{-11}
Capacitance	C	cm	$\text{Farad} = \dfrac{\text{Coulomb}}{\text{Volt}}$	$\dfrac{1}{9 \times 10^{11}}$	$\dfrac{1}{9 \times 10^{11}}$
Charge	Q, q	$\text{cm}^{\frac{3}{2}} \text{g}^{\frac{1}{2}} \text{sec}^{-1}$	Coulomb	$\dfrac{1}{3 \times 10^9}$	$\dfrac{1}{3 \times 10^9}$
Current	I, i	$\text{cm}^{\frac{3}{2}} \text{g}^{\frac{1}{2}} \text{sec}^{-2}$	$\text{Ampere} = \dfrac{\text{Coulomb}}{\text{sec}}$	$\dfrac{1}{3 \times 10^9}$	$\dfrac{1}{3 \times 10^9}$
Dielectric displacement Polarization Surface charge density	D P σ, ρ	$\text{cm}^{-\frac{1}{2}} \text{g}^{\frac{1}{2}} \text{sec}^{-1}$	$\dfrac{\text{Coulomb}}{\text{length}^2}$	$\dfrac{1}{4\pi \times 3 \times 10^9}$ $\dfrac{1}{3 \times 10^9}$	$\dfrac{1}{4\pi \times 3 \times 10^5}$ $\dfrac{1}{3 \times 10^5}$
Current density Potential decay function	J, j $\Psi(t)$	$\text{cm}^{-\frac{1}{2}} \text{g}^{\frac{1}{2}} \text{sec}^{-2}$	$\dfrac{\text{Ampere}}{\text{length}^2}$	$\dfrac{1}{3 \times 10^9}$	$\dfrac{1}{3 \times 10^5}$
Electromotive force or Potential difference	V	$\text{cm}^{\frac{1}{2}} \text{g}^{\frac{1}{2}} \text{sec}^{-1}$	Volt	300	300

I. UNITS AND CONSTANTS

Field	E, F	$cm^{-\frac{1}{2}} g^{\frac{1}{2}} sec^{-1}$	$\dfrac{\text{Volt}}{\text{length}}$	300	3×10^4
Resistance	R	$cm^{-1} sec$	$\text{Ohm} = \dfrac{\text{Volt}}{\text{Ampere}}$	9×10^{11}	9×10^{11}
Resistivity	ρ	sec	Ohm . length	9×10^{11}	9×10^9
Conductivity†	γ, σ, κ	sec^{-1}	Ohm^{-1} . length^{-1}	$\dfrac{1}{9 \times 10^{11}}$	$\dfrac{1}{9 \times 10^9}$
Current decay function	$\dot{\Psi}(t)$				

† The symbol γ in the monograph denotes conductivity in practical units

APPENDIX II

Formulae for the Capacitance and Equivalent Circuits

For an empty plate capacitor the capacitance in the electrostatic system is

$$C_0 = \frac{\text{area}}{4\pi d} \qquad (\text{II.1})$$

where d is the distance between plates, all distances being measured in centimetres. When the capacitance is required in Farads

$$C = 4\pi\varepsilon_0 C_0. \qquad (\text{II.2})$$

These formulae neglect the fringing of the field at the edges of the capacitor plates.

When a capacitor is filled with a dielectric its static capacitance is, in practical units,

$$C = 4\pi\varepsilon_0\varepsilon_s C_0 \qquad (\text{II.3})$$

fringe fields being neglected. Allowance for fringe fields, i.e. for the electrical field caused by the charges on the capacitor in the space outside it, may be made according to formulae given in references 7 and 8 of Chapter 6. The capacitance under dynamic conditions may be expressed in terms of $\varepsilon^*(\omega)$, measured in e.s.u. The complex capacitance defined by equation (1.15) is thus

$$C^*(\omega) = 4\pi\varepsilon_0 C_0 \varepsilon^*(\omega). \qquad (\text{II.4})$$

Alternating current measurements give the impedance $Z^*(\omega)$ of the capacitor containing the dielectric which is related to the complex capacitance by equation (1.22). The impedance can be represented by an equivalent circuit. For a given frequency ω an impedance can be represented by a large number of possible equivalent circuits, the simplest of which are the parallel and the series circuit discussed in Chapter 1. Table I compares the characteristics of these circuits for a given ω.

If the same dielectric is represented in one case by the parallel and in the other by the series circuit

$$\tan\delta = \frac{1}{\omega C_p R_p} = \omega C_s R_s \qquad (\text{II.5})$$

while the values of the equivalent series and parallel components are related by

$$\frac{C_p}{C_s} = \frac{1}{1 + \tan^2 \delta} \tag{II.6}$$

and

$$\frac{R_p}{R_s} = \frac{\tan^2 \delta + 1}{\tan^2 \delta}. \tag{II.7}$$

In the parallel representation it is convenient to work with the resistivity of a material which is measured as the resistance between electrodes applied to

TABLE C

Circuit	$Z^*(\omega)$	$\|Z(\omega)\|$	$C^*(\omega) = \dfrac{1}{i\omega Z^*(\omega)}$	$\tan \delta$	$\tan \phi$	τ
R_s —C_s (series)	$R_s - \dfrac{i}{\omega C_s}$	$\dfrac{1}{\omega C_s}\sqrt{1 + \omega^2 C_s^2 R_s^2}$	$\dfrac{C_s}{1 + i\omega C_s R_s}$	$\omega C_s R_s$	$\dfrac{1}{\omega C_s R_s}$	$C_s R_s$
$R_p \parallel C_p$	$\dfrac{R_p}{1 + i\omega C_p R_p}$	$\dfrac{R_p}{\sqrt{1 + \omega^2 C_p^2 R_p^2}}$	$C_p - \dfrac{i}{\omega R_p}$	$\dfrac{1}{\omega C_p R_p}$	$\omega C_p R_p$	$\dfrac{1}{\omega^2 C_p R_p}$

opposite faces of a cube of unit edge. A capacitor of these dimensions has in the electrostatic system a capacitance C_0 equal $1/4\pi$ centimetres so that equation (11.4) gives

$$\varepsilon^*(\omega) = \frac{1}{\varepsilon_0} C^*(\omega) \tag{II.8}$$

or, in the parallel representation,

$$\varepsilon^*(\omega) = \frac{1}{\varepsilon_0}\left(C_p(\omega) - \frac{i}{\omega R_p(\omega)}\right) \tag{II.9}$$

$$= \frac{1}{\varepsilon_0}\left(C_p(\omega) - \frac{i}{\omega}\gamma(\omega)\right)$$

where $\gamma(\omega)$ is the conductivity in the practical system used, C_p the capacitance in the same system, and $\varepsilon^*(\omega)$ is in e.s.u.

Apart from the simple circuits described above and in Chapter 1, the most

II. FORMULAE FOR THE CAPACITANCE AND EQUIVALENT CIRCUITS

useful equivalent circuit is the two-layer model, pictured inset in Fig. 1. This circuit and its generalization to n parallel circuits in series is treated in some detail by J. Volger (1960. *Prog. Semicond.*, **4**, 209).

The two-layer model represents two layers of dielectric of thickness d_1 and d_2, static dielectric constant ε_1 and ε_2 and direct conductivity γ_1 and γ_2.

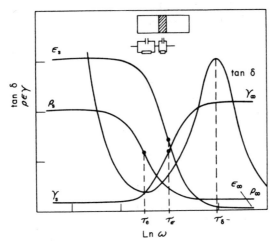

FIG. 1. Example of calculated curves giving the dependence of permittivity ε, conductivity γ, resistivity ρ, and tan δ on ln ω in the two-layer model. Arbitrary units. From the position of the curves along the frequency axis, the relevant relaxation times may be derived.

Volger evaluates the frequency dependent behaviour of this model in terms of the following formulae

$$\varepsilon'(\omega) = \frac{\varepsilon_s + \varepsilon_\infty \tau_\varepsilon^2 \omega^2}{1 + \tau_\varepsilon^2 \omega^2} \tag{II.10}$$

$$\gamma'(\omega) = \frac{\gamma_s + \gamma_\infty \tau_\gamma^2 \omega^2}{1 + \tau_\gamma^2 \omega^2} \tag{II.11}$$

$$\rho'(\omega) = \frac{\rho_s + \rho_\infty \tau_\rho^2 \omega^2}{1 + \tau_\rho^2 \omega^2} \tag{II.12}$$

$$\tan \delta = \frac{\gamma_\infty}{\varepsilon_0 \omega \varepsilon_\infty} \frac{\left(\frac{\varepsilon_\infty \gamma_s}{\varepsilon_s \gamma_\infty}\right) + \tau_\delta^2 \omega^2}{1 + \tau_\delta^2 \omega^2} \tag{II.13}$$

with

$$\tau_\varepsilon = \tau_\gamma = \varepsilon_0 \frac{\varepsilon_s - \varepsilon_\infty}{\gamma_\infty - \gamma_s} = \varepsilon_0 \frac{\varepsilon_1 d_2 + \varepsilon_2 d_1}{\gamma_1 d_2 + \gamma_2 d_1} \tag{II.14}$$

$$\tau_\rho = \left(\frac{\gamma_\infty}{\gamma_s}\right)^{\frac{1}{2}} . \tau_\varepsilon \qquad (II.15)$$

$$\tau_\delta = \left(\frac{\varepsilon_\infty}{\varepsilon_s}\right)^{\frac{1}{2}} . \tau_\varepsilon \qquad (II.16)$$

$$\varepsilon_s = \frac{(d_1 + d_2)}{\left(\frac{d_1}{\gamma_1} + \frac{d_2}{\gamma_2}\right)^2} . \left(\frac{d_1 \varepsilon_1}{\gamma_1^2} + \frac{d_2 \varepsilon_2}{\gamma_2^2}\right) \qquad (II.17)$$

$$\varepsilon_\infty = \frac{d_1 + d_2}{\dfrac{d_1}{\varepsilon_1} + \dfrac{d_2}{\varepsilon_2}} \qquad (II.18)$$

$$\gamma_s = \frac{d_1 + d_2}{\dfrac{d_1}{\gamma_1} + \dfrac{d_2}{\gamma_2}} \qquad (II.19)$$

$$\gamma_\infty = \frac{d_1 + d_2}{\left(\dfrac{d_1}{\varepsilon_1} + \dfrac{d_2}{\varepsilon_2}\right)^2} \left(\frac{d_1 \gamma_1}{\varepsilon_1^2} + \frac{d_2 \gamma_2}{\varepsilon_2^2}\right) \qquad (II.20)$$

Equations (II.17)–(II.20) give the static respectively high frequency values for the composite dielectric, while the significance of the three relaxation times are indicated in Fig. 1. The conductivities are in practical units, the dielectrics constants in e.s.u. Volger evaluates these formulae further for special cases.

A number of other equivalent circuits, some including inductances, are evaluated by B. B. East and W. B. Westphal (May 1964. Technical Report 189, Laboratory for Insulation Research, Massachusetts Institute of Technology). These authors define the dielectric constant in practical units, so that the dimensionless $\varepsilon^*(\omega)$ appears as $\dfrac{\varepsilon^*}{\varepsilon_0}$.

APPENDIX III

The Limiting Values of ε_S and tan δ for a given Number of Dipoles

The relationship between ε_s and the number of dipoles N in a dielectric is not straightforward because of uncertainties about electrostatic interaction on a microscopic scale (see Chapter 8). The relationship between the maximum loss angle observed and ε_s is also not straightforward because relaxation times are usually distributed more or less widely. However, the question may arise as to whether dielectric relaxation is a sufficiently sensitive method for the detection of a suspected number N of dipoles. In that case it is useful to know an upper limit for tan δ. Such a limit follows from equation (8.16) if a single relaxation time is assumed in addition to the other, unrealistic, assumptions inherent in that equation. Table D gives the upper limit for ε_s as well as tan δ for dipoles with $\mu = 2 \times 10^{-18}$ e.s.u. as a function of their number N cm^{-3}. Two temperatures have been chosen to give respectively $kT = 4 \times 10^{-15}$ and $kT = 4 \times 10^{-14}$. The two higher values chosen for ε_∞ refer to unusual conditions and are of interest only for materials with ferroelectric possibilities (see Chapter 16).

TABLE D

$T = 28.9°K$	$T = 289°K$		$\varepsilon_\infty = 2$		$\varepsilon_\infty = 10$		$\varepsilon_\infty = 100$		$\varepsilon_\infty = 1000$	
N	N	$N\mu^2/kT$	ε_s	$\tan \delta_m$	ε_s	$\tan \delta_m$	ε_s	$\tan \delta_m$	ε_s	$\tan \delta_m$
10^{15}	10^{16}	10^{-6}	2·00	$<10^{-4}$	10·00	$<10^{-4}$	100·02	$1·0 \times 10^{-4}$	1001·41	$7·0 \times 10^{-4}$
10^{16}	10^{17}	10^{-5}	2·00	$<10^{-4}$	10·00$_2$	$1·0 \times 10^{-4}$	100·15	$7·5 \times 10^{-4}$	1014·22	$7·1 \times 10^{-3}$
10^{17}	10^{18}	10^{-4}	2·00$_2$	$5·6 \times 10^{-4}$	10·02	$1·0 \times 10^{-3}$	101·20	$5·9 \times 10^{-3}$	1163·00	$>10^{-1}$
10^{18}	10^{19}	10^{-3}	2·02$_2$	$5·5 \times 10^{-3}$	10·20	$1·0 \times 10^{-2}$	116·94	$3·2 \times 10^{-2}$		
10^{19}	10^{20}	10^{-2}	2·23$_7$	$5·6 \times 10^{-2}$	12·42	$1·1 \times 10^{-1}$				
10^{20}	10^{21}	10^{-1}	7·0$_6$	$>10^{-1}$						

APPENDIX IV

The Four Forms of the Complete Duhamel Integral

Equation (5.4c) is a special formulation of the Duhamel integral (J. M. C. Duhamel. 1833. *J. Ec. polytech.*, Vol. 22). Four equivalent forms of the complete integral may be derived as follows.

Assume a sudden stepwise increase of the stimulus from zero to ΔE at time 0. The response to this step at a time t_2 is given by

$$\Delta E \cdot \phi(t_2).$$

Here

$$\Psi(t_2) = \phi(0) - \phi(t_2) \tag{IV.1}$$

in terms of the notation of Chapter 5, where the potential decay function was defined with reference to a discharge, i.e. to a negative ΔE.

If instead of being stepwise the stimulus varies continuously with time then

$$\Delta E = \dot{E}\,\Delta t = \frac{dE}{dt}\Delta t$$

will be the contribution to E of the small time interval Δt and the response at time t_2 to all changes in the past of E will be

$$\sum_i \dot{E}(t_i)\,\Delta_i t\, \phi(t_2 - t_i) \to \int_{-\infty}^{t_2} \frac{dE(t)}{dt}\phi(t_2 - t)\,dt.$$

The changes of E do not necessarily determine E completely and therefore in order to obtain the total response $r(t_2)$ in all cases we have to add a term taking account of the initial value (although in most cases its contribution will be zero). Thus (using the notation of Chapter 5, with t for t_2 and u for t)

$$r(t) = E(-\infty)\phi(\infty) + \int_{-\infty}^{t} \frac{dE(u)}{dt}\phi(t - u)\,du. \tag{A}$$

Applying partial integration this can be written

$$r(t) = E(-\infty)\phi(\infty) + \left|E(u)\phi(t-u)\right|_{-\infty}^{t} - \int_{-\infty}^{t} E(u)\frac{d\phi(t-u)}{du}\,du.$$

$$\tag{IV.2}$$

The function under the integral may be written

$$\frac{d\phi(t-u)}{du} = -\dot{\phi}(t-u)$$

$$= \dot{\Psi}(t-u)$$

in view of equation (IV.1). Evaluation of equation (IV.2) now gives

$$r(t) = E(t)\phi(0) + \int_{-\infty}^{t} E(u)\dot{\phi}(t-u)\,du \tag{B}$$

which is a more general form of equation (5.4c), and equivalent to equation (6.4). The term before the integral in equation (B) is zero if $\phi(0)$ is derived from equations (5.1) which do not take the instantaneous response into account. More generally, if the instantaneous response is included in $r(t)$, the constant term is given by $\varepsilon_\infty E(t)$.

By a change of variable we can write A and B as

$$r(t) = E(-\infty)\phi(\infty) + \int_0^\infty \dot{E}(t-x)\phi(x)\,dx \tag{C}$$

and

$$r(t) = E(t)\phi(0) + \int_0^\infty E(t-x)\dot{\phi}(x)\,dx. \tag{D}$$

Form (D) is equivalent to equation (6.5). If there is no stimulus before $t = 0$ the limits of integration for equations (A)–(D) are from zero to t.

Equations (A)–(D) are quoted in reference 4 of Chapter 5. The equivalence of the four forms shows that the distinction between a polarization decay function and a current decay function is not important or even necessarily useful in a purely mathematical context. It is important only where physical dimensions are considered.

APPENDIX V

Some General Expressions Relating to Distribution Functions

The stimulus $s(t)$ which figures in a linear differential equation may be of different types, for example periodic, stepwise, or a delta function. The integral expressions which may be derived for a given differential equation and various types of stimuli are all interrelated. Many investigators express this interrelation in terms of Laplace transforms (see references 2 and 4 of Chapter 5). The relationship between the current decay function $\dot{\Psi}(t)$ and the generalized Debye equations (5.13) and (5.14) reads, in terms of Laplace transforms (Macdonald and Barlow. 1963. Rev. mod. Phys., 35, 940),

$$Q(p) = \int_0^\infty \dot{\Psi}(t) \, e^{-pt} \, dt \tag{V.1}$$

$$\equiv L(\dot{\Psi}(t)) \tag{V.2}$$

$$= \int_0^\infty \frac{G(\tau) \, d\tau}{1 + p\tau} \tag{V.3}$$

The variable p is complex and for the purposes of the present application can best be interpreted as $p = i\omega$. The function $G(\tau)$ in this expression is

$$G(\tau) = \frac{y(\tau)}{\varepsilon_s - \varepsilon_\infty} \tag{V.4}$$

since for the purposes of superposition integrals the integral over the distribution function (see equation 5.11) is usually taken to be unity. Equations (6.21) and (6.22) are a special case of the above general relations.

The Laplace transform and the functions defined above appear in the more mathematical literature of dielectric relaxation in various ways.

$$\dot{\Psi}(t) \equiv L^{-1}(Q(p)) \tag{V.5}$$

defines the inverse Laplace transform. The integral in equation (V.3) is frequently expressed in terms of a variable

$$s = \log_e \frac{\tau}{\tau_0} \tag{V.6}$$

where τ_0 is a positive constant. Now a distribution function $F(\tau)$ or $F(s)$ is defined by

$$F\left(\log \frac{\tau}{\tau_0}\right) \equiv \tau G(\tau) \tag{V.7}$$

where

$$\int_{-\infty}^{+\infty} F(s)\,ds = \int_0^\infty G(\tau)\,d\tau = 1 \tag{V.8}$$

while

$$Q(p) = \int_{-\infty}^{\infty} \frac{F(\tau)}{1 + p\tau}\,ds. \tag{V.9}$$

If real and imaginary parts are separated in equations (V.3) or (V.9) and we put $p = i\omega$

$$Q(i\omega) = J(\omega) - iH(\omega) \tag{V.10}$$

$$= \frac{1}{\varepsilon_s - \varepsilon_\infty}(\varepsilon'(\omega) - i\varepsilon''(\omega)) \tag{V.11}$$

where

$$J(\omega) = \int_0^\infty \frac{G(\tau)\,d\tau}{1 + (\omega\tau)^2} \tag{V.12}$$

$$= \int_0^\infty \frac{F(s)\,ds}{1 + (\omega\tau)^2}$$

while

$$H(\omega) = \int_0^\infty \frac{\omega\tau G(\tau)\,d\tau}{1 + (\omega\tau)^2} \tag{V.13}$$

$$= \int_0^\infty \frac{\omega\tau F(s)\,ds}{1 + (\omega\tau)^2}.$$

Those of the above relations which involve τ may be alternatively expressed in terms of

$$z \equiv \frac{\tau_0}{\tau} = e^{-s} \tag{V.14}$$

where τ_0 is constant.

Laplace and other integral transforms are treated in textbooks. The following may be useful:

V. EXPRESSIONS RELATING TO DISTRIBUTION FUNCTIONS

H. S. Carslaw and J. C. Jaeger (1947). "Operational Methods of Applied Mathematics." Oxford University Press, London.

G. Doetsch (1961). "Guide to Applications of Laplace Transforms." Van Nostrand, New York.

N. W. McLachlan (1963). "Complex Variable Theory and Transform Calculus." Cambridge University Press, London.

APPENDIX VI

Abstracts and Collections of Data

National Academy of Sciences, National Research Council, Washington, D.C., U.S.A. (1966). "Digest of the Literature on Dielectrics." Annual Publication. Vol. 30.

von Hippel (editor) (1954). "Dielectric Materials and Applications." John Wiley, New York.

F. M. Clark (1962). "Insulating Materials for Design and Engineering Practice." John Wiley, New York.

Landolt Börnstein (1957). "Numerical Data and Functional Relationships in Physics, Chemistry, Astronomy, Geophysics and Technology." Vol. IV, Part 3. Springer-Verlag, Berlin.

F. Buckley and A. A. Maryott (1958). "Table of Dielectric Dispersion Data for Pure Liquids and Dilute Solutions." National Bureau of Standards, U.S.A., Circular 589.

Plastics Materials Guide (1966). Originally published with February 1966 issue of "Plastics", Heywood, London, in co-operation with I.C.I. Plastics Division.

Author Index

Numbers in parentheses are reference numbers. Numbers in italics indicate the page on which the references are listed.

Abadie, P., 230 (32) *231*
Abraham, M., 1 (1) 2 (1) *12*, 13 (1) *31*
Agranovskaya, A. I., 251 (22) *254*
American Society for Testing and Materials, 87 (9) 92 (9) *93*
Anderson, P. W., 139 (9) *142*
Andrade, E. N. da C., 150 (15) *162*
Auty, R. P., 223 (18) *231*
Axilrod, B. M., 175 (22) *183*

Ballantyne, J. M., 249 (20) *254*
Bardeen, J., 213 (15) *216*
Barker, A. S., 249 (19) *254*
Barlow, C. A., Jr., 69 (4) 71 (4) *77*
Barlow, H. M., 92 (12) *93*
Barrie, I. T., 188 (9) *201*
Beckenridge, R. G., 176 (23) *183*
Becker, R., 1 (1) 2 (1) *12*, 13 (1) *31*
Bell, D. A., 189 (12) *201*
Bendat, J., 53 (9) 56 (9) 57 (9) 58 (9) 59 (9) *64*
Ben Sira, M. Y., 214 (17) 215 (17) *216*
Bergen, J. T., 69 (3) 71 (3) *77*
Bernal, J. D., 222 (14) *231*
Birks, J. B., 161 (40) *163*, 187 (4) *201*
Birnbaum, G., 136 (5, 6, 7) 137 (5, 6) 138 (6) 140 (5, 6) 141 (5) 142 (11) *142*
Bjerrum, N., 223 (25) *231*
Bleaney, B., 131 (8) *132*
Bogdanov, S. V., 234 (6) 251 (27) 252 (27) *253, 254*
Boltzmann, L., 76 (9) *77*
Born, M., 128 (7) *132*
Böttcher, C. J. F., 19 (2) 28 (2) 30 (2) *31*, 111 (2) 114 (2) *121*
Brophy, J. J., 243 (13) 244 (13) 250 (13) *254*
Brot, C., 159 (30) *162*
Brown, J., 203 (1) 208 (1) *216*

Buchanan, T. J., 156 (27) 157 (27) *162*
Buckingham, K. A., 188 (9) *201*
Bueche, F., 193 (19) *201*
Bunn, C. W., 168 (10) 169 (10) *182*
Burgess, R. E., 51 (5) 53 (8) *64*, 238 (10, 11) 240 (10) *253*
Burnstein, E., 131 (13) *133*

Chan, R. K., 228 (27) *231*
Charbonniere, R., 230 (32) *231*
Church, H. F., 229 (31) 230 (33) *231*
Clemett, C., 166 (5) 167 (7) 172 (7) *182*
Cochran, W., 247 (16) 248 (16) 249 (16) *254*
Cole, K. S., 83 (2) *93*, 95 (1) 96 (1) 102 (1) 103 (5) *109*
Cole, R. H., 83 (2) *93*, 95 (1) 96 (1) 102 (1) 103 (5, 8) 104 (9, 10) *109*, 118 (11) 121 (18) *121*, *122*, 160 (34) 161 (39) *163*, 223 (18) *231*
Cook, H. F., 156 (27) 157 (27), *162*
Cook, J. S., 171 (15a) *183*
Coulson, C. A., 155 (23) *162*
Cowley, R. A., 248 (18) 249 (18) *254*
Cross, L. E., 251 (24, 25) 252 (29) 253 (29) *254*
Cullen, A. L., 92 (12) *93*
Curtis, A. J., 185 (1) 187 (1) *200*

Daniel, V., 168 (9) 169 (9) 171 (9) 172 (9, 16, 16a) *182, 183*, 208 (9) 215 (20) *216*, 218 (4) 220 (7, 8, 9) 221 (4, 7, 8, 11) *231*
Davidson, D. W., 104 (9) *109*, 160 (34) *163*, 228 (27) *231*
Davies, M., 166 (5) 167 (7) 172 (7) *182*
Davies, R. O., 36 (6) *45*, 195 (20) *201*
Day, A. G., 229 (30) *231*

Debye, P., 24 (4) *31*, 114 (5) 121 (5) *121*, 145 (4) 152 (4) *162*
de Groot, S. R., 37 (4) *45*
Dekker, A. J., 179 (30) *183*
Delahay, P., 213 (13) 214 (13) *216*
de Loor, G. P., 208 (7) *216*
de Maeyer, L., 158 (29) *162*, 223 (23, 24) 224 (23, 24) 225 (23, 24) 228 (23, 24) *231*
Denbigh, K. G., 33 (1) 40 (1) 43 (1) *45*
Denney, D. J., 159 (32) 161 (37) *163*
Devonshire, A. F., 233 (2) 236 (2) 237 (9) 238 (9) 240 (9) 246 (2) *253*
Diamant, H., 235 (8) *253*
di Marzio, E. A., 197 (22) *201*
Drenk, K., 235 (8) *253*
Dryden, J. S., 168 (8) 171 (15) 172 (15) 173 (18) 177 (24) *182*, *183*, 208 (6) *216*
Dunning, W. J., 166 (2) *182*
Dupuis, M., 222 (15, 16) 223 (16) 226 (15, 16) 227 (16) *231*

Earley, J. E., 161 (39) *163*
Eigen, M., 157 (28) 158 (28, 29) *162*, 223 (23, 24) 224 (23, 24) 225 (23, 24) 228 (23, 24) *231*
Eitel, W., 199 (25) *201*
Electrical Research Association, 87 (8) 91 (8) 92 (8) *93*
Estin, A., 136 (6) 137 (6) 138 (6) 140 (6) *142*
Ewald, P. P., 146 (5) 151 (5) *162*
Eyring, H., 61 (12, 13) *64*, 167 (6) *182*

Ferry, J. D., 188 (7) 191 (7) 193 (7) 194 (7) 196 (7) *201*
Field, S. B., 131 (12) *133*
Fischer, E., 148 (11) 149 (11) *162*
Forsbergh, P. W., 135 (4) *142*
Fowler, R. H., 33 (2) 36 (2) 37 (2) 41 (2) *45*, 47 (1) 61 (1) *64*, 222 (14) *231*
Frenkel, L., 142 (12) *142*
Fröhlich, H., 19 (3) 20 (3) *31*, 34 (3) 38 (3) *45*, 52 (6) *64*, 75 (6) 77, 99 (4) 100 (4) 101 (4) *109*, 111 (3) 114 (3) 115 (3) 116 (3) 117 (3) 118 (8) *121*, 127 (5) *132*, 170 (14) 181 (33) 182 (34) *182*, *183*, 237 (12) 242 (12) 254 (12) *254*
Fuoss, R. M., 103 (7) *109*, 194 (17) *201*
Fürth, R., 59 (11) *64*

Garton, C. G., 187 (5) 189 (13) *201*, 208 (10) 209 (10) 210 (10) 213 (10) *216*, 230 (33) *231*
Genzel, L., 131 (14) *133*
Gibbs, J. H., 197 (22) *201*
Gidel, A., 230 (32) *231*
Gierer, A., 149 (12) *162*
Gilchrist, A., 161 (39) *163*
Girard, P., 230 (32) *231*
Glarum, S. H., 119 (13) 120 (13) *121*, 159 (31) *163*
Glasstone, S., 61 (12) *64*, 167 (6) *182*
Gottlieb, M., 131 (9) *132*
Gränicher, H., 222 (17) 223 (21, 22) 224 (22) 225 (17, 21, 22) *231*
Grant, E. H., 156 (27) 157 (27) *162*
Grant, F. A., 97 (2) 98 (2) *109*
Gross, B., 9 (3) *12*, 69 (2) 71 (2) 73 (2) 75 (2) *76*
Gross, E. P., 125 (3) 128 (3) *132*, 139 (10) *142*
Gubkin, A. N., 251 (26) *254*
Guggenheim, E. A., 33 (2) 36 (2) 37 (2) 41 (2) *45*, 47 (1) 61 (1) *64*
Guilbot, A., 230 (32) *231*
Guillemin, E. A., 1 (2) 2 (2) 3 (2) 9 (2) 10 (2) *12*, 43 (7) 44 (7) *45*
Gurney, R. W., 179 (29) 180 (29) *183*

Hague, B., 87 (10) *93*
Hamon, B. V., 85 (3) *93*, 218 (2) *231*
Happ, H., 131 (14) *133*
Harnik, E., 214 (17) 215 (17) *216*
Hass, M., 131 (13) *133*
Hasted, J. B., 155 (21) 156 (21) 158 (21) *162*
Heller, W. R., 177 (27) *183*
Henvis, B. W., 131 (13) *133*
Herzfeld, K. F., 36 (5) 41 (5) 42 (5) *45*, 135 (1) 140 (1) *142*, 144 (3) 145 (3) 150 (3) 160 (3) *162*
Hill, N., 150 (14) 151 (17) 156 (26) 157 (26) *162*
Hill, R. M., 245 (15) 250 (15) *254*
Hoffmann, J. D., 175 (21, 22) *183*
Hopkinson, J., 69 (1) 71 (1) 76 (1) *76*, 79 (1) 83 (1) 85 (1) *93*
Huang, K., 128 (7) *132*
Humbel, F., 223 (19) *231*

AUTHOR INDEX

Hydeshima, T., 188 (8) 191 (8) 192 (8) 194 (8) 195 (8) 196 (8) *201*

Ichiki, S. K., 245 (15) 250 (15) *254*
Illinger, K. H., 128 (6) *132*, 135 (2) 137 (2) 142 (2) *142*, 143 (2) 146 (2) 148 (2) 149 (2) 151 (2) 154 (2) 157 (2) *162*
Isupov, V. A., 251 (22, 23) 253 (23) *254*
Iwayanagi, S., 188 (8) 191 (8) 192 (8) 194 (8) 195 (8) 196 (8) *201*

Jaccard, C., 222 (17) 225 (17) *231*
Jackson, W., 207 (5) *216*
Jacobs, P. W. M., 215 (19) *216*
Jona, F., 223 (19) *231*, 233 (1) 236 (1) 237 (1) 240 (1) 241 (1) 242 (1) 246 (1) 247 (1) 250 (1) *253*
Jones, G. G., 131 (10) *133*
Jones, G. O., 195 (20) *201*

Kail, J. A. E., 221 (10) *231*
Känzig, W., 233 (4) 236 (4) *253*
Kashtanova, A. M., 251 (26) *254*
Kats, A., 200 (30) *201*
Kharadly, M. M. Z., 207 (5) *216*
Kirkwood, J. G., 103 (7) *109*, 116 (7) *121*, 148 (9) 156 (25) *162*, 193 (17) *201*
Kiseleva, K. V., 251 (27) 252 (27) *254*
Kittel, C., 49 (3) 53 (3) 55 (3) 59 (3) *64*, 76 (7) *77*
Kneser, H. O., 223 (20) 228 (20) *231*
Kondrat'ev, V. N., 61 (14) *64*
Kono, R., 160 (36c) *163*
Krasucki, Z., 211 (12) *216*, 230 (33) *231*
Kröger, F. A., 182 (37) *183*, 215 (22) *216*
Kubo, R., 119 (14) 121 (14) *122*
Kurtze, G., 157 (28) 158 (28) *162*

Laidler, K. J., 61 (12) *64*, 167 (6) *182*
Lamb, J., 36 (6) *45*
Lauritzen, J. I., 173 (19) 174 (19) 175 (19) *183*
Lax, M., 50 (4) *64*
Le Fèvre, R. J. W., 149 (13) *162*
Lidiard, A. B., 180 (31) *183*
Litovitz, T. A., 36 (5) 41 (5) 42 (5) *45*, 135 (1) 140 (1) *142*, 144 (3) 145 (3) 150 (3) 160 (3, 35, 36a, 36b, 36c) 161 (35) *162*, *163*
Lorentz, H. A., 125 (2) *132*

Loubser, J. H. N., 131 (8) *132*
Lurio, A., 244 (14) *254*

McClellan, A. L., 155 (24) 157 (24) 162 (24) *162*, 229 (29) 230 (29) *231*
MacDonald, J. R., 69 (4) 71 (4) 77, 105 (11) *109*
McDuffie, G. E., 160 (35, 36a, 36b, 36c) 161 (35) *163*
Magun, S., 223 (20) 228 (20) *231*
Many, A., 214 (17) 215 (17) *216*
Martin, D. H., 131 (10, 12) *133*
Maryott, A. A., 136 (6, 7) 136 (6) 138 (6) 140 (6) 142 (11) *142*
Matveeva, N., 251 (21) *254*
Maxwell, J. C., 76 (13) *77*, 204 (2) *216*
Maycock, J. N., 215 (19) *216*
Meakins, R. J., 146 (6) 147 (6) 151 (6) 152 (6) *162*, 168 (8) 170 (13) 171 (13) 173 (13) 174 (20) 175 (20) 177 (13, 24) *182*, *183*, 208 (6) *216*, 218 (1, 2, 3) 219 (1) 221 (3) *231*
Megaw, H. D., 233 (5) 235 (5) *253*
Merz, W. J., 233 (3) 236 (3) 244 (3) 250 (3) *253*
Miller, R. C., 153 (18) *162*
Mopsik, F. J., 104 (10) *109*
Morin, F. J., 182 (35) *183*
Mott, N. F., 179 (29) 180 (29) *183*
Müller, A., 168 (11) 169 (12) 174 (12) *182*

Nowick, A. S., 177 (27) *183*, 214 (18) *216*

O'Dwyer, J. J., 153 (19) 154 (19) *162*
Okano, K., 188 (8) 191 (8) 192 (8) 194 (8) 195 (8) 196 (8) *201*
Oldham, J. W. H., 218 (4) 221 (4) *231*
Olson, H. F., 12 (4) *12*, 30 (5) *31*
Onsager, L., 115 (6) *121*, 222 (15, 16) 223 (16) 226 (15, 16) 227 (16) *231*
Oster, G., 156 (25) *162*
Owston, P. J., 221 (12) *231*

Palik, E. D., 131 (11) *133*
Parker, R. A., 214 (16) *216*
Pauling, L., 155 (22) *162*, 222 (13) 229 (13) *231*
Pelzer, H., 9 (3) *12*, 62 (15) *64*, 190 (14) *201*
Pepinski, R., 235 (8) *253*

Perls, T. A., 161 (38) *163*
Perrin, F. J., 148 (10) *162*
Pfeiffer, H. G., 175 (21) *183*
Phillips, C. S. E., 135 (3) *142*
Pimentel, G. C., 229 (29) 230 (29) *231*
Pincus, G., 131 (13) *133*
Pinnentel, G. C., 155 (24) 157 (24) 162 (24) *162*
Pitt, D. A., 147 (8) 151 (8, 16) *162*
Popov, S. N., 251 (22) *254*
Pöschl, Th., 146 (5) 151 (5) *162*
Powles, J. G., 118 (12) 119 (12) *121*, 221 (10) *231*
Prandtl, L., 146 (5) 151 (5) *162*
Pratt, B., 214 (17) 215 (17) *216*
Price, W. C., 166 (3) *182*

Quinn, R. G., 160 (35) 161 (35) *163*

Rassushin, V. A., 251 (27) 252 (27) *254*
Read, B. E., 106 (12) *109*
Reddish, W., 87 (5, 6) *93*, 188 (9) 197 (23, 24) 198 (23) 199 (23) *201*
Rice, S. O., 56 (10) 57 (10) 58 (10) *64*
Rogers, M. G., 215 (20) *216*
Rouse, P. E., 193 (18) 194 (18) *201*

Sack, R. A., 153 (19) 154 (19) *162*, 218 (1) 219 (1, 5, 6) 221 (6) 224 (5) *231*
Saito, K., 188 (8) 191 (8) 195 (8) 196 (8) *201*
Sautter, D., 189 (11) *201*
Sawyer, C. B., 235 (7) *253*
Scaife, B. K. P., 52 (7) *64*, 98 (3) 99 (3) *109*, 118 (9, 10) 112 (15) 121 (16) *121*, *122*, 154 (20) *162*
Schatz, E. R., 182 (38) *183*
Scheiber, D. J., 88 (11) 92 (11) *93*
Scherrer, P., 222 (17) 223 (19) 225 (17) *231*
Sewell, G. L., 181 (33) *183*
Sheridan, J., 124 (1) *132*
Shirane, G., 233 (1) 236 (1) 237 (1) 240 (1) 241 (1) 242 (1) 246 (1) 247 (1) 250 (1) *253*
Sillars, R. W., 73 (5) *77*, 204 (4) 207 (4) *216*
Skanavi, G. I., 251 (21, 26) 252 (28) *254*
Smolenski, G. A., 251 (22) *254*

Smyth, C. P., 143 (1) 147 (8) 151 (1, 8, 16) 153 (18) 159 (33) *162*, *163*, 166 (4) *182*
Spatz, H. C., 223 (24) 224 (24) 225 (24) 228 (24) *231*
Spenke, E., 210 (11) 211 (11) 212 (11) *216*, 226 (26) *231*
Spernol, A., 149 (12) *162*
Staveley, L. A. K., 173 (17) *183*
Steinemann, A., 222 (17) 225 (17) *231*
Stern, E., 244 (14) *254*
Stevels, J. M., 177 (28) 178 (28) *183*, 200 (26, 27, 29) *201*, 213 (14) 214 (14) *216*
Stratton, J. A., 111 (1) *121*
Sullivan, E. P. A., 149 (13) *162*
Sussmann, J. A., 228 (28) *231*
Sutter, P. H., 214 (18) *216*
Sutton, P. M., 200 (28) *201*
Szigeti, B., 113 (4) *121*, 188 (10) *201*, 247 (17) 248 (17) *254*

Tamm, K., 157 (28) 158 (28) *162*
Temperley, H. N. V., 193 (16) *201*, 208 (8) *216*
Thurn, H., 191 (15) *201*
Timmermans, J., 165 (1) 173 (1) *182*
Tinkham, M., 249 (19) *254*
Tobolski, V., 188 (6) 191 (6) 193 (6) *201*
Toll, J. S., 76 (8) *77*
Tolman, R. C., 47 (2) 49 (2) 53 (2) *64*
Tower, C. H., 235 (7) *253*
Turner, C., 172 (16a) *183*

Urry, D. W., 61 (13) *64*

Vallauri, M. G., 135 (4) *142*
van Amerongen, C., 213 (14) 214 (14) *216*
van Houten, S., 182 (36) *183*
van Vleck, J. H., 127 (4) 128 (4) *132*
Vein, P. R., 220 (7) 221 (7) *231*
Volger, J., 177 (25, 28) 178 (28) 182 (25) *183*, 200 (29) *201*, 213 (14) 214 (14) 215 (21) *216*
von Hippel, A. R., 87 (7) *93*, 185 (2) 187 (2, 3) *200*
von Schweidler, E., 76 (11) *77*

AUTHOR INDEX

Wachtman, J. B., 177 (26) *183*
Wagner, K. W., 76 (12) *77*, 204 (3) 206 (3) *216*
Wasilik, J. H., 214 (16) *216*
Weber, R., 131 (14) *133*
Weisskopf, V. F., 127 (4) 128 (4) *132*, 137 (8) 141 (8) *142*
Whalley, E., 228 (27) *231*
Wiechert, E., 76 (10) *77*
Wigner, E., 62 (15) *64*

Wilkinson, G. R., 166 (3) *182*
Williams, G., 86 (4) 87 (4) *93*, 103 (6) 106 (12) *109*, 147 (7) *162*, 196 (21) *201*
Wilner, L. B., 161 (38) *163*
Wirtz, K., 149 (12) *162*
Wolf, K., 191 (15) *201*

Ziegler, G., 223 (20) 228 (20) *231*
Ziman, J. M., 181 (32) *183*
Zwanzig, R., 121 (17) *122*

Subject Index

Absorption
  dielectric (see relaxation and resonance)
  infrared or optical, 123, 128, 223, 247
  ultrasonic (see ultrasonic relaxation)
Activation energy, 22, 61
Alcohols, 158, 217
Amorphous solids (see glasses and polymers)
Anharmonic vibrations, 249
Arc plots
  Cole–Cole (permittivity), 95
  Grant (conductivity), 97
  Scaife (polarizability), 98
Aroclor, 161
Associated liquids, 154, 193

Barrier layer, 214
Bistable model (see potential well models)
Boltzmann constant, 255
Boltzmann distribution, statistics, 48, 180, 212
Bridge measurements, 87
Brownian motion, 27

Capacitor, capacitance, 1, 13, 256, 259
  edge effects, 259
Cellulose, 187, 229
Ceramics, 200, 215
Circuits, equivalent or analogous, 1, 30, 73, 74, 214, 259
Clausius–Mossotti equation, 28, 114
Clusters in liquids, 154, 160
Cole–Cole plot (see arc plots)
Collisions
  in gases, 123, 127, 139
  in liquids, 150
Conductivity
  direct, ohmic, steady, 81, 212, 236, 257
  electronic, 181
  high frequency, 19
  ionic, 179, 212
  plot, 97
Continuity equation, 212
Cooperative phenomena, 20
Correlation function (autocorrelation function), 53
Curie point (see ferroelectric transition), 236

Damping (see resonance)
Debye equations, 18
Debye unit, 255
Decay functions, 68, 80, 120, 159, 257
Defect effects, 176, 215, 250
Deformation losses (in glasses), 200
Dielectric
  artificial, 208
  displacement, 14
  loss, 18
  material, ix
Dielectric constant (see permittivity)
  complex, 13, 15
  incremental, 213, 235
  low field, 238
  optical, 16
  static, 1, 16
Diffusion (Fick's equation), 26, 179
Dipole, 20
  bistable, 20
  contribution to cohesive energy, 166
  moment, 20
Dispersion, dielectric (see frequency dependence of dielectric constant)
  anomalous, 126
Dissipation of energy, 42
Dissociation equilibrium, 36, 157, 179, 224
Distribution functions, 70, 99, 188, 267
Domain, ferroelectric, 233
Domain wall movement, 250
Double layer (see barrier layer)
Double well (see potential well models)

SUBJECT INDEX

Einstein–Nernst relation, 180
Electrode effects, 211
Electronic effects, 181, 252
Electrostatic interaction, 27, 39, 111
Entropy, 33, 48
  generation of (see dissipation of energy), 42
Equivalent-circuits (see circuits)
Esters, 171

Feedback, positive, 30
Field
  cavity, 112, 115
  electrical, 13
  internal, 27
  reaction, 115
Fluctuations, 47
  measurements of, 243
  rate of, 53
Free energy, 34
  Gibbs, 192, 237
  Helmholtz, 34

Garton effect, 211
Gaussian distribution, 51
  random process, 56
Glass transition (see transition)
Glassy state, 159, 194
Glycerol, 159

Heat, specific, 242
Hydrogen bonding, 155, 187, 217, 221, 228, 245
Hysteresis, 234

Ice, 221
Impedance, 5
Inductance, 10
Injection of carriers, 212
Insulating materials (see dielectric), ix
  definition of practical limits, 211
  deterioration of, 215, 230
  low loss, 188

Kramers–Kronig relations, 73, 76

Laplace transforms, 120, 267
Lattice dynamics, 128, 247
Life time, 210, 226
Linearity of response, definition, 66

Liquids, 143
  associated, 154
  Kneser (group 2), 145
Loss tangent, 5, 18
  lowest limit for, 188

Markoff process, 56
Mass action, law of, 210, 224
Maxwell–Wagner effect, 203
Methanes, substituted, 136, 154, 165
Mobility, drift, 181
Mode of relaxation, 193
Moisture (see water)

Noise
  definition, 53
  measurement, 243
  $1/f$ noise, 188

Ohmic electrodes, 82, 211
Optical constants, 129

Paper capacitors, 161, 211, 230
Permittivity (see dielectric constant)
Perovskite structure, 246
Poisson's equation, 212
Polarizability, 28
  of sphere, 112
Polarization, 15
  dipolar, 16
  electronic (see optical)
  interfacial, 82
  optical, 16
Polyethylene, 188
Polymers, organic, 185
Polyvinyl chloride, 197
Potential well models
  double well (bistable model), 19, 36, 43, 49, 56, 179, 192, 241
  multiple wells
    chains of, 219
    in planar array, 175
    in three dimensions, 177, 221
Power spectrum (see noise)
Pre-exponential factor, 63, 105

Random processes, 53, 149, 225
Rate constants, 62, 221
Rate processes, discussion, 61
Recombination of ions, 208, 221

Refractive index, 16, 128
Regression time (see relaxation time)
Relaxation, ix
  differential equation of, 7, 15, 24
Relaxation time
  distributed (see distribution functions)
  pressure dependence of, 42, 161, 228
  single, 3, 13
    extrinsic, 29, 119, 153
    intrinsic, 29, 119, 153
    of bistable model, 24, 56
    of Debye model, 27
Resistivity (see conductivity)
Resonance, 11, 123

Saturation, 23, 233
Sawyer and Tower circuit, 234
Schering Bridge, 87
Space charge, 213
Space group, polar, 233
Spectral density (see noise)
Step-function methods of measurement
  (see direct current), 79
Stoichiometry, 182, 215
Susceptibility, 238
Symmetric top molecules (see methanes, substituted)

Tensors, 237
Terylene, 197
Thermal motion, 21, 47
Thermodynamics, 33
  irreversible, 39
Time constant (see relaxation time)
Transformer ratio bridges, 89
Transitions
  ferroelectric, 234
  glass, 194
  order–disorder, 36, 236
  phase, 165
Triglycine sulphate, 241

Ultrasonic relaxation, 42, 135, 154, 157, 160, 177, 182, 185, 223
Uncertainty relation, 123
Units and conversion factors, 255

Vibrating reed voltmeter, 79
Vibration losses, 200
Viscosity, 27, 146, 150, 153
Volume, free, available, 154, 167, 196

Water, 155, 221
  in insulation, 228
Wiener–Khintchine theorem, 55